W9-AVL-801

FOSS Science Resources

Earth and Sun

Full Option Science System
Developed at
The Lawrence Hall of Science,
University of California, Berkeley
Published and distributed by
Delta Education,
a member of the School Specialty Family

1487713
978-1-62571-372-8
Printing 1 – 7/2015
Quad/Graphics, Versailles, KY

Table of Contents

Table of Contents (continued)

Investigation 5: Water Planet

References

Changing Shadows

Objects you can't see through, such as people, birds, buildings, balls, and flagpoles, have **shadows** on sunny **days**. That's because opaque objects block sunlight. A shadow is the dark area behind an **opaque** object. Shadows give information about the position of the **Sun**. If you see your shadow in front of you, then the Sun is behind you.

Did you know that a shadow tells you what time of day it is? Let's see how that works. First, if you are in North America, you need to be facing south. When you are facing south, north is behind you.

It is 12:00 noon. Let's look at the flagpole and observe its shadow. What direction is the shadow pointing?

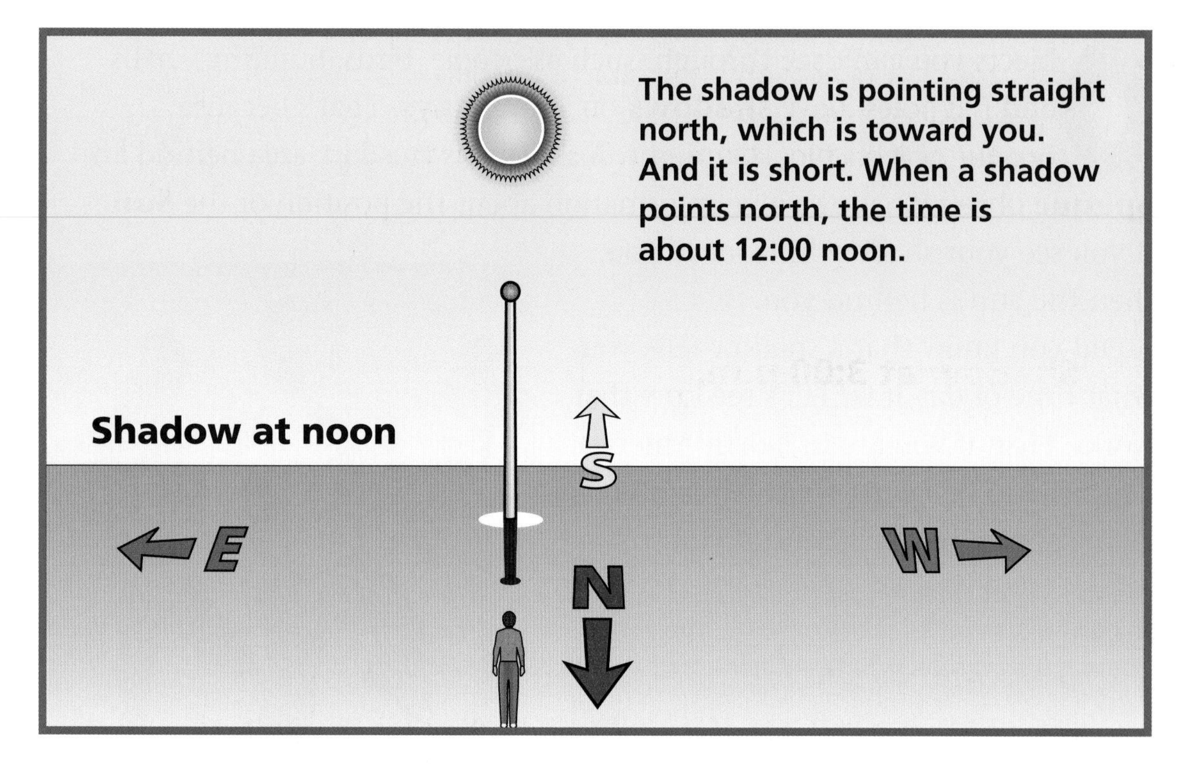

What does the flagpole's shadow look like at 9:00 in the morning? Did you see your shadow this morning? Do you remember what direction it was pointing? Do you remember how long it was?

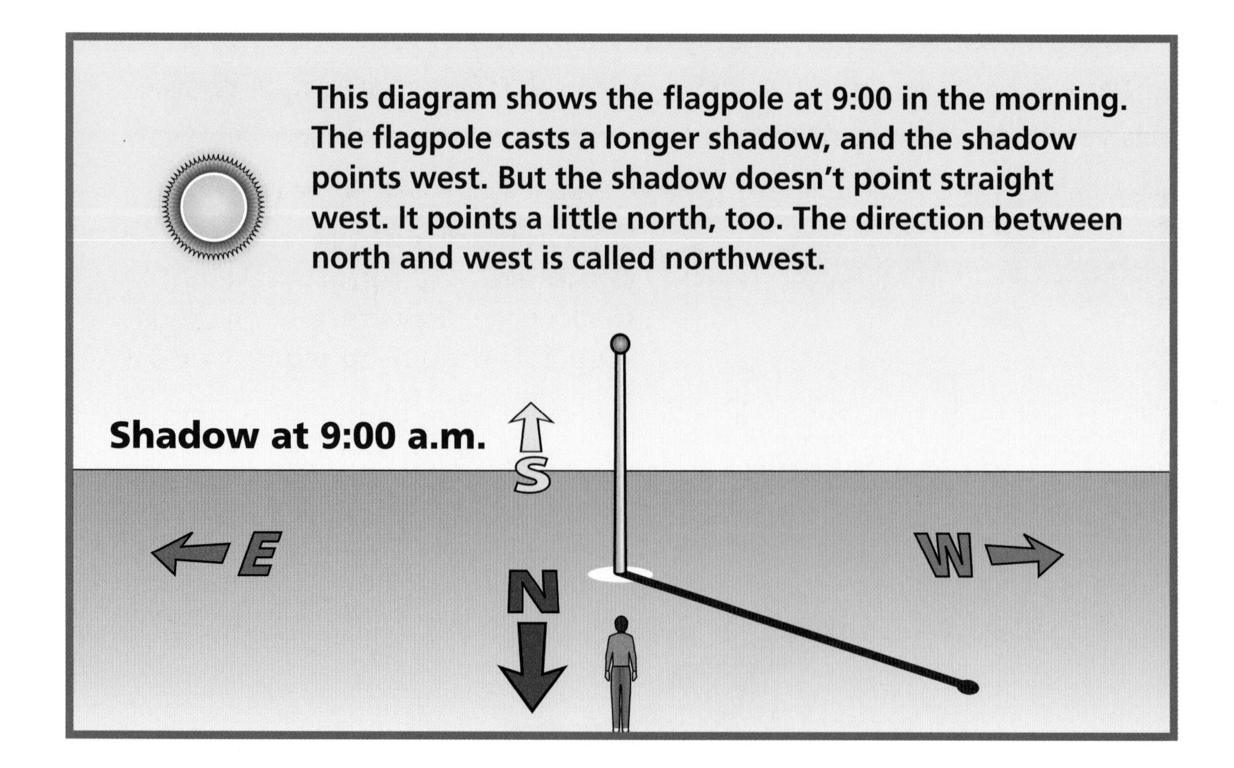

What happens to the shadow in the afternoon?

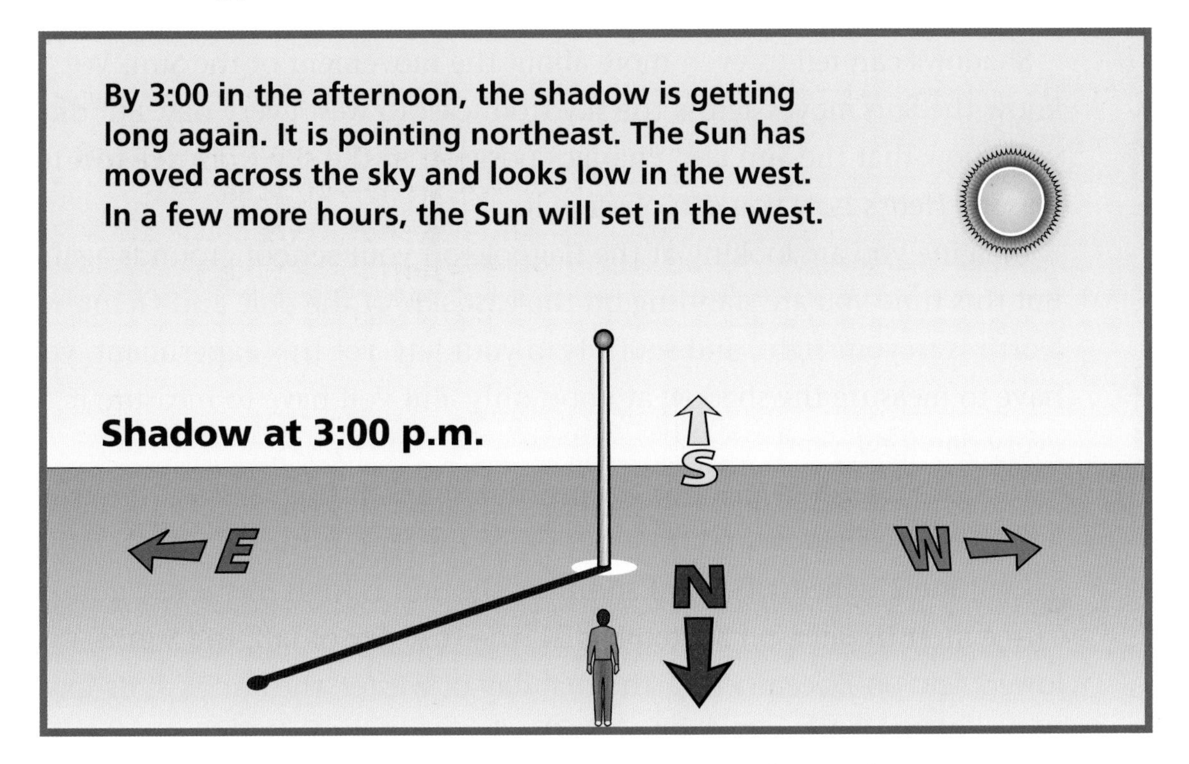

Two things happen to a shadow between **sunrise** and **sunset**: its length changes and its direction changes. Early in the morning, a shadow is long, and it points west. We observe that the Sun moves across the sky from east to west, and the shadow changes.

At noon, the Sun reaches its highest point in the sky. Now the shadow is as short as it will get. It points straight north. We observe that the Sun keeps moving across the sky. Just before sunset, a shadow is very long, and it points east.

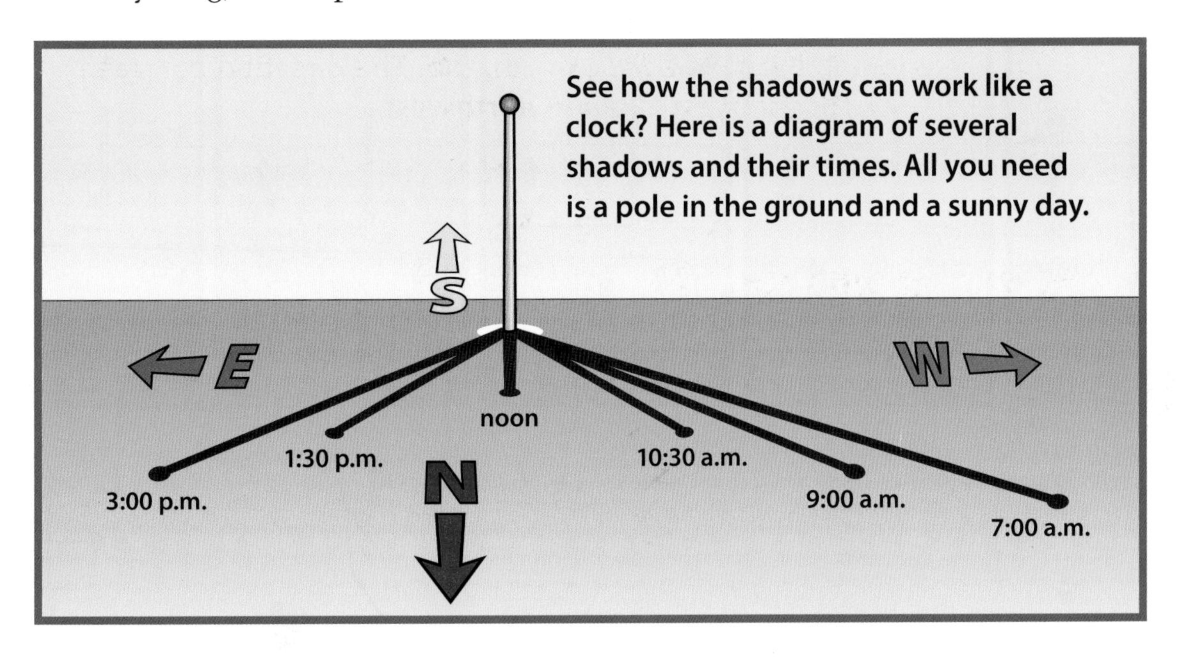

The Sun and Seasons

Shadows can tell us even more about the movement of the Sun. We know the Sun moves across the sky from east to west every day. But did you know that the Sun also changes position in the sky from **season** to season? Here's how you can observe it.

Imagine you are looking at the flagpole on your school grounds again. But this time you are standing on the east side of the pole and facing west. North is to your right, and south is to your left. For this experiment, you have to measure the shadow at noon only. But you have to measure it every day for 1 year!

Here are the noon shadows for just five times during the year. Look at the length of the shadow and the position of the Sun in the sky on each date.

On June 21, the first day of summer, the Sun is high in the sky at noon. Three months later, on September 21, the first day of fall, the Sun is lower. And on December 21, the first day of winter, the Sun is at its lowest noon position. After December 21, the Sun begins to climb higher in the sky again. On March 21, the first day of spring, it is as high as it was in September. One year after starting the experiment, on June 21, the Sun is again at its highest noon position.

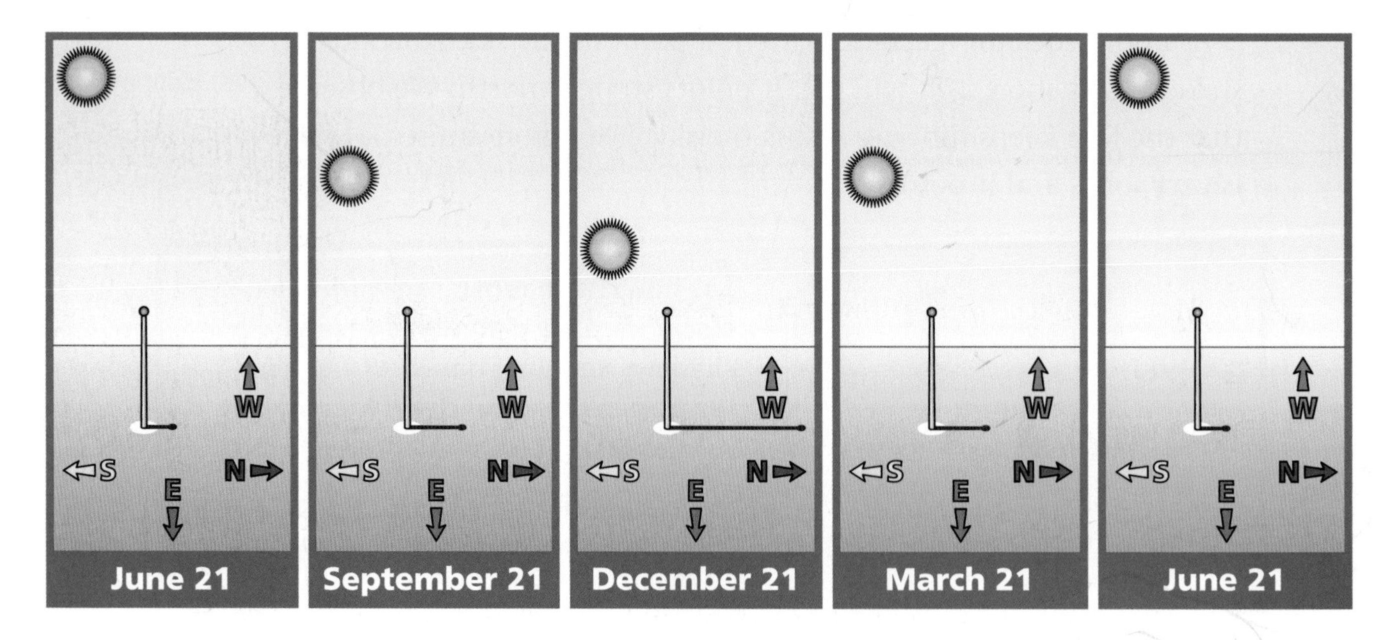

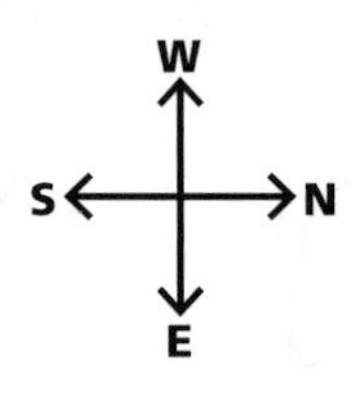

The Sun's change of position in the sky minute by minute during a day is predictable. The Sun's position in the sky season to season during a year is also predictable.

Thinking about Shadows

1. How does the Sun's position in the sky change over 1 day?
2. In what ways do shadows change during the day?
3. What causes shadows to change during the day?
4. Think about a flagpole. How does its shadow change over 1 year?
5. Look at the photo at the top of the page. Can you see the shadow of the person? Can you see the shadows of the four flagpoles? Why or why not?

The Sun rising over a cornfield in Minnesota

Sunrise and Sunset

The Sun has just come up in this picture. It is sunrise. What direction are you looking?

The Sun always rises in the east. If you are in Portland, Maine, the Sun rises in the east. If you are in Portland, Oregon, the Sun rises in the east. If you are in Raleigh, North Carolina, the Sun rises in the east. If you are in Brownsville, Texas, or Broken Bow, Oklahoma, the Sun rises in the east. Wherever you are on **Earth**, the Sun rises in the east.

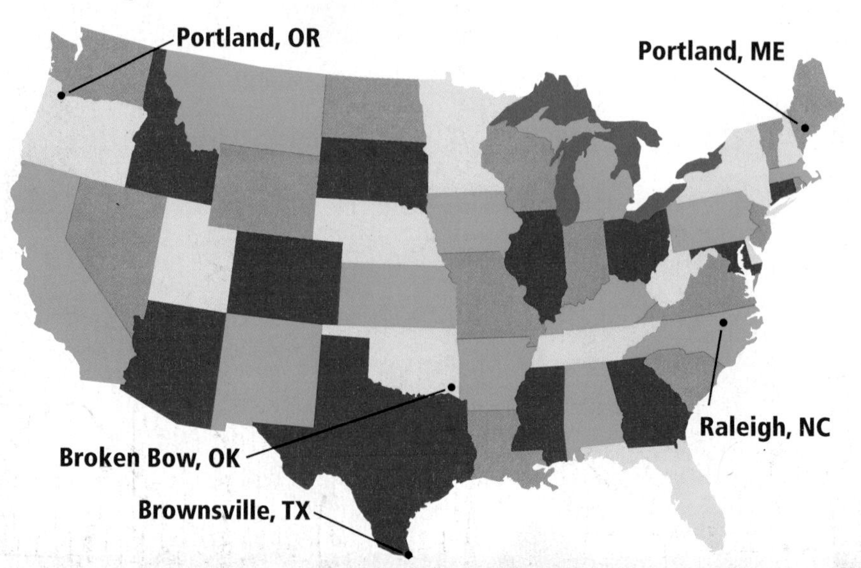

In this picture, the Sun is just about to go down. It is sunset. What direction are you looking now?

That's right, you're looking west. The Sun always sets in the west. If you are in Portland, Maine, the Sun sets in the west. If you are in Portland, Oregon, the Sun sets in the west. If you are in Raleigh, North Carolina, the Sun sets in the west. If you are in Brownsville, Texas, or Broken Bow, Oklahoma, the Sun sets in the west. Wherever you are on Earth, the Sun sets in the west.

Every day the Sun rises in the east and sets in the west. To get from east to west, the Sun appears to slowly travel across the sky. In the early morning, when the Sun first comes up, it is touching the horizon in the east. At noon, the Sun is at its highest position in the sky. At sunset, the Sun is touching the horizon in the west. The Sun's position in the sky changes all day long.

The Sun setting over the city of Boston, Massachusetts

There is one thing you can depend on for sure. The Sun will come up tomorrow morning. And you can be sure it will come up in the east. When the Sun is in the sky, you can feel its warmth. At the end of the day, it will set in the west. You can count on it.

As the day goes along, it looks as though the Sun travels across the sky from east to west. During the morning, it rises higher and higher in the sky. At noon, it is at its highest position in the sky. From noon to sunset, the Sun continues to travel west. And it gets lower and lower in the sky. At sunset, the Sun disappears below the horizon in the west. Another day has passed. And tomorrow will be the same.

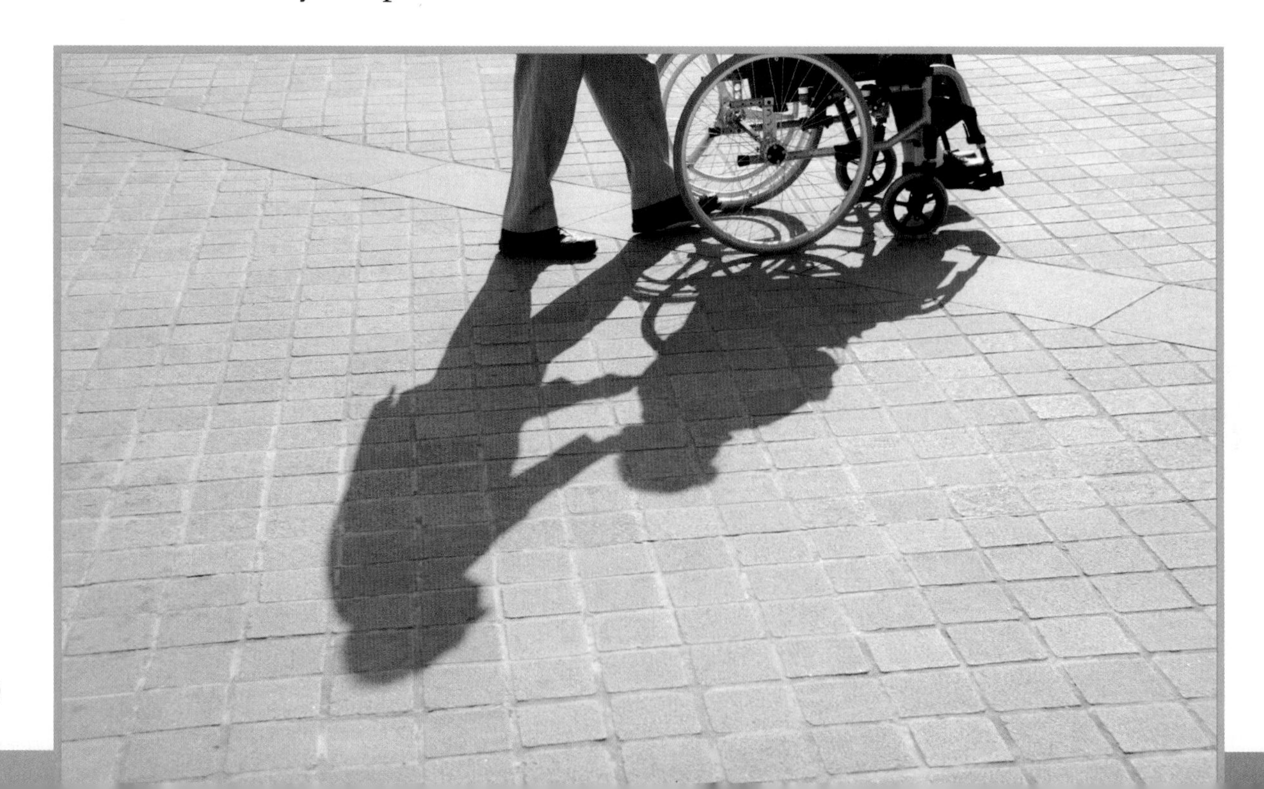

Earth's Rotation

The Sun looks as though it moves across the sky. But it really doesn't. It is Earth that is moving. Here's how it works.

Earth is spinning like a top. It takes 1 day (24 hours) for Earth to **rotate** once. Because Earth is rotating, half of the time we are on the sunny side of Earth. We call the sunny side day. The other half of the time we are on the dark side of Earth. We call the dark side **night**.

Imagine it's just before sunrise. You can't see the Sun because you are still on the dark side of Earth. But in 5 minutes, Earth will rotate just enough for you to see the Sun come over the horizon. That moment is sunrise.

Earth turns toward the east, the direction of the orange arrow. That means the first sunlight of the day will be in the east. And, of course, Earth keeps turning. You keep moving with it. In 4 or 5 hours, you have turned so far that the Sun is high over your head. And 5 hours after that, the Sun is low in the western sky. This is because Earth is moving in an eastward direction. It looks as though the Sun is moving across the sky in a westward direction. Finally, it is sunset. The Sun slips below the horizon in the west. It is dark again.

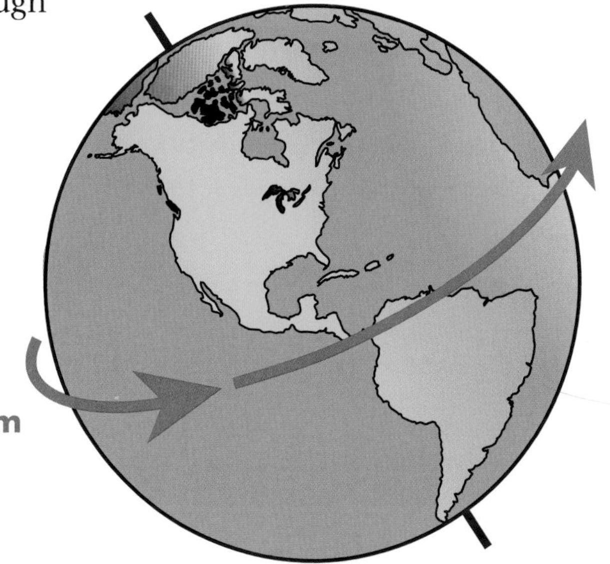

Earth turns toward the east. So the Sun seems to move from east to west across the sky.

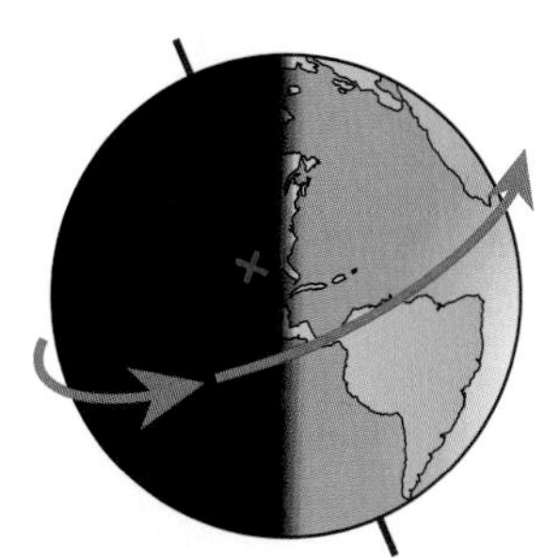

The x shows your position just before sunrise.

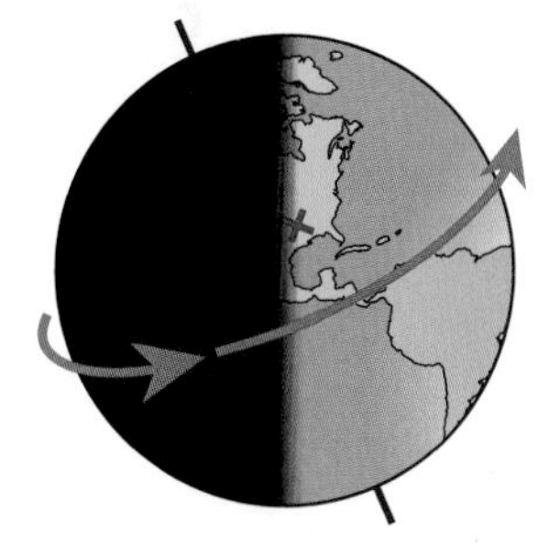

The x shows your position just after sunrise.

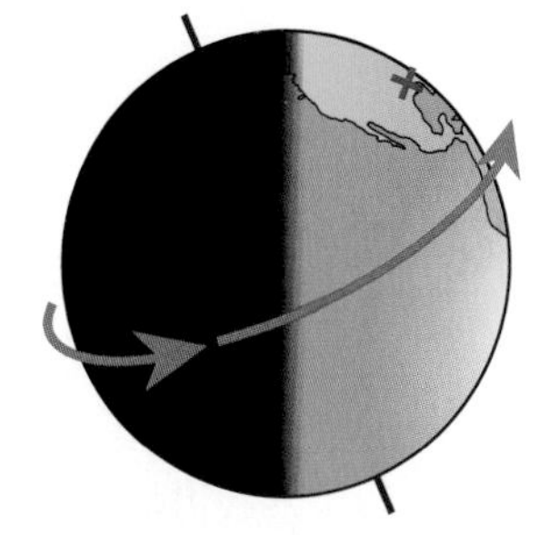

The x shows your position near noon.

Shadows

A shadow is the dark area behind an opaque object. It is created where an object blocks sunlight. A steel pole, like a flagpole, casts a shadow. The direction of the pole's shadow changes as the Sun's position changes. At noon, the Sun is highest in the sky. Noon is also when the flagpole's shadow is the shortest of the day.

We can watch the noon shadow to see how the Sun's position changes from season to season. The length of that shadow changes a little bit every day. Why does the length of the shadow change? It changes because the position of the Sun at noon changes a little bit every day.

The Sun's position changes all day from sunrise to sunset.

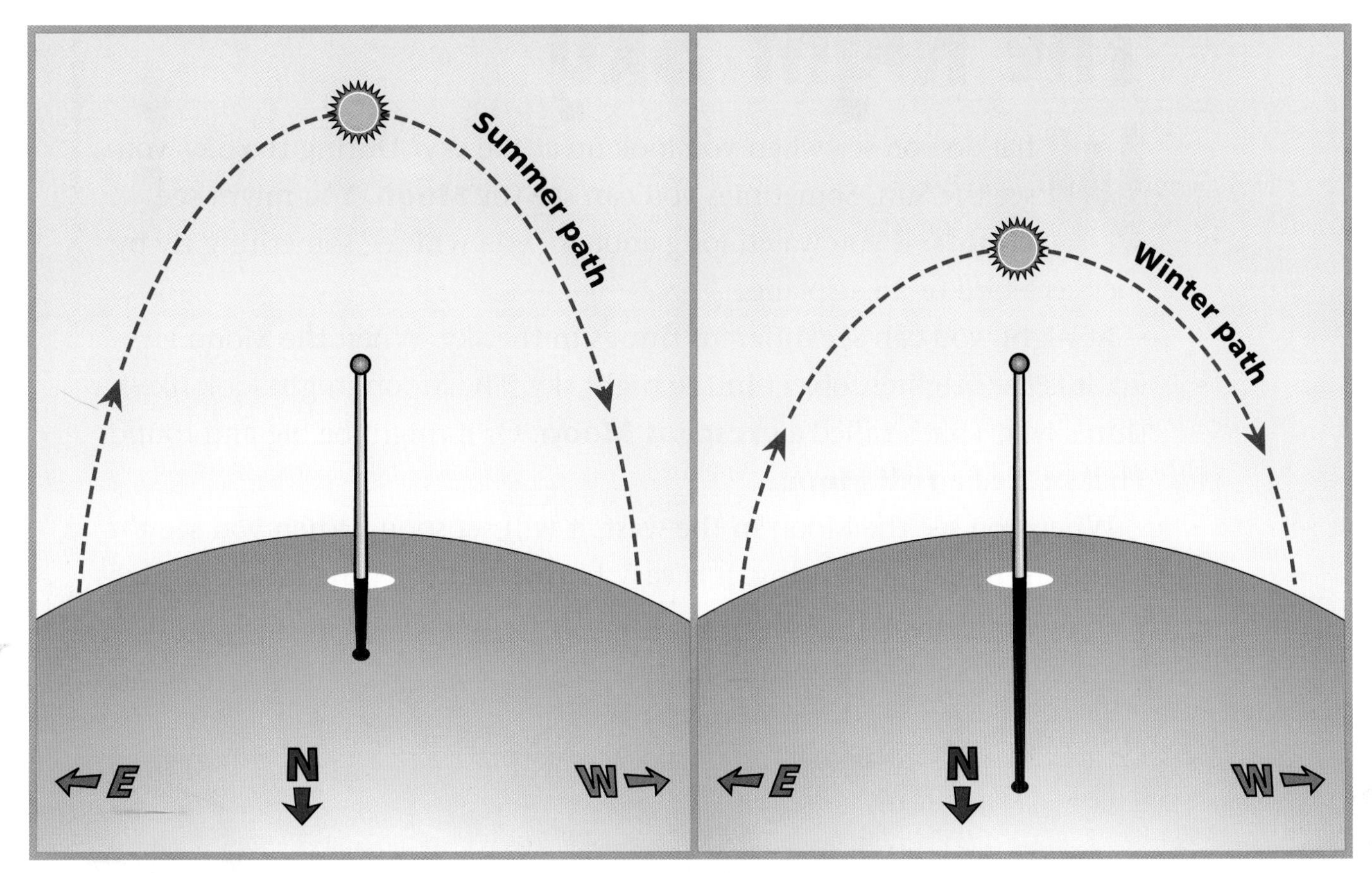

The Sun's path through the sky is higher in summer.

The pattern of change is predictable. In North America, the position of the noon Sun gets higher in the sky from December 21 to June 21. On June 21, the Sun is highest in the sky. That's also the day when the flagpole's shadow is the shortest of the year.

The position of the noon Sun gets lower in the sky each day between June 21 and December 21. On December 21, the noon Sun is lowest in the sky. That's also the day that the flagpole's shadow is the longest of the year.

The Sun's position in the sky changes in two ways. Every day the Sun rises in the east, appears to travel across the sky, and sets in the west. The other way the Sun's position changes is in its daily path. In summer, the Sun's path is high in the sky. In winter, the Sun's path is lower in the sky.

The Night Sky

What do you see when you look up at the sky? During the day, you see the Sun. Sometimes you can see the **Moon**. You might see **clouds**. If you watch long enough, you will see something fly by, such as a bird or an airplane.

At night, you can see different things in the sky. When the Moon is up, it is the brightest object in the night sky. The Moon might look like a thin sliver. That's called a **crescent Moon**. Or it might be big and round. That's called a **full Moon**.

When you see the Moon in the west, it will set soon. When you see the Moon in the east, it is rising. It is easy to **predict** the time of day or night the Sun will rise and set. It is much harder to predict the time of day or night the Moon will rise and set.

A full Moon over New York City

A crescent Moon

The Moon during the day

On a clear night, you can see about 2,000 stars in the sky.

Stars

When it is clear, you can see **stars** in the night sky. Night is the only time you can see stars. Well, almost the only time. There is one star we can see in the daytime. It's the Sun. The Sun shines so brightly that it is impossible to see the other stars. But after the Sun sets, we can see that the sky is full of stars. It looks like there are millions of stars in the sky on a clear night. But actually you can see only about 2,000 stars with your **unaided eyes**.

Venus and Jupiter in the eastern sky just before sunrise

Planets

Some stars are brighter than others. They are the first ones you can see just after the Sun sets. Did you ever make a wish on the first star that appears in the evening sky? "Star light, star bright, first star I see tonight. I wish I may, I wish I might, have the wish I make tonight." That star might not be a star at all. The brightest stars are actually **planets**. That's one way you can tell a planet from a star, by how brightly it appears to shine.

Earth **orbits** the Sun with seven other planets and several **dwarf planets**. You can see five planets in the night sky. Venus is one of the planets you might see. Ancient sky watchers called Venus the evening star. It is seen near the western horizon after sunset. Venus was also called the morning star. It is also seen near the eastern horizon just before sunrise. What caused the confusion?

Two planets orbit closer to the Sun than Earth does. Mercury is closest to the Sun. Then comes Venus. Venus takes only 225 days to go around the Sun. Sometimes Venus is positioned where we can see it from Earth just before sunrise as the morning star. A few months later, Venus has traveled to the other side of the Sun. Now it is positioned for us to see it after sunset as the evening star. That's why ancient sky watchers thought Venus was two different stars.

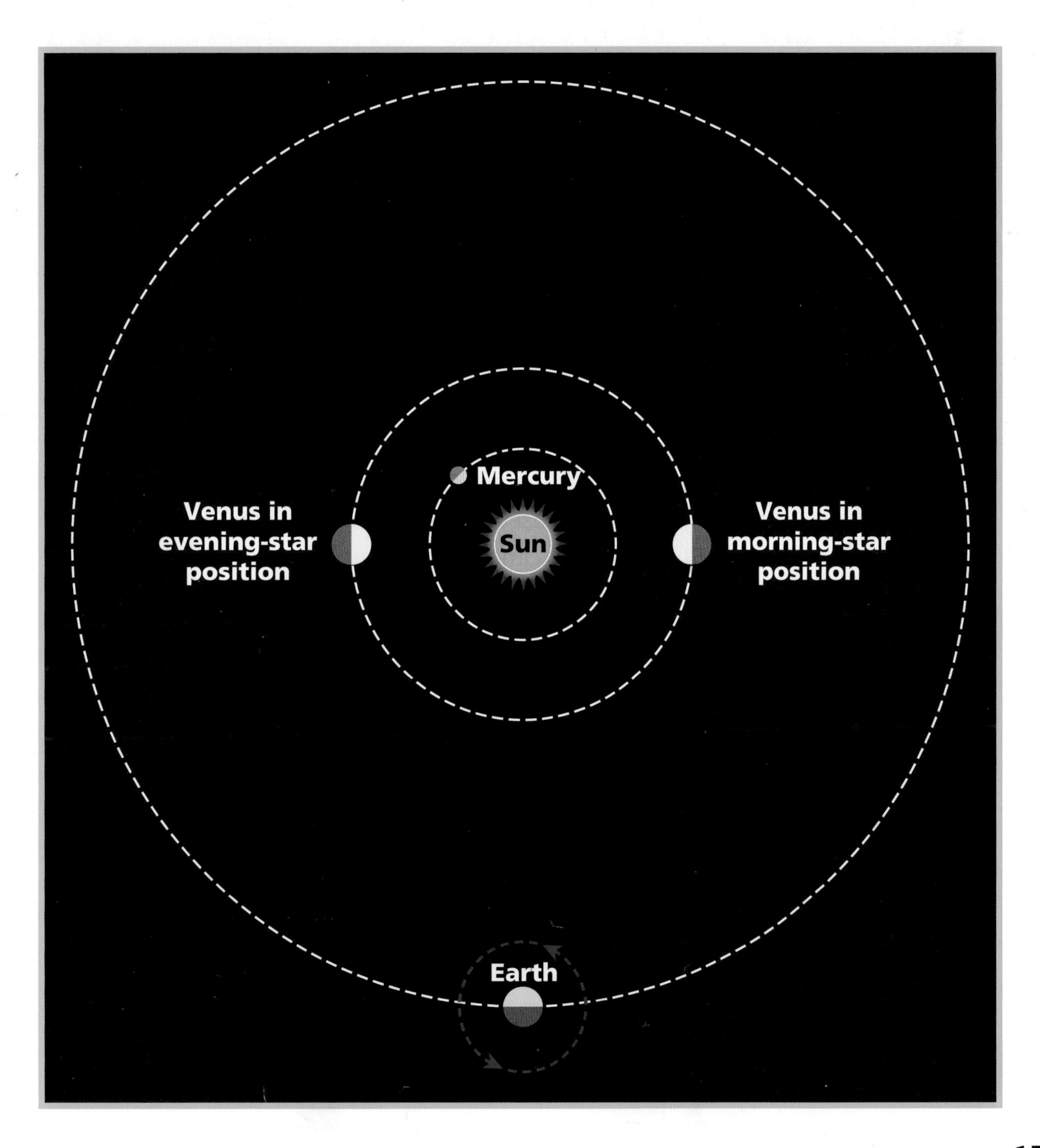

Venus is often visible from Earth.

Four other planets can be seen with unaided eyes. Mercury is visible sometimes. Because it is so close to the Sun, it is often lost in the bright glare of the Sun. Mars is the fourth planet from the Sun. It shines with a slightly red light. Jupiter and Saturn are the farthest of the visible planets. Still, they are pretty bright because they are so big.

It is a special night when you can see all five planets at the same time in the night sky. It doesn't happen very often. It happened in 2004. It won't happen again until 2036!

Thinking about the Night Sky

1. What are some of the objects you can see in the night sky that you can't see in the day sky?
2. Which object is the brightest object in the night sky?
3. Which star is the closest to Earth?
4. Look at the picture of the crescent Moon. What is the other bright object you can see in the night sky?

Looking through Telescopes

Galileo Galilei

What do you see when you look at the sky on a clear night? You probably see many twinkling stars. Maybe you see the Moon or a planet. People saw the same objects in the sky thousands of years ago.

The way we look at objects in the sky changed in 1608. In that year, the **telescope** was invented. A telescope is a tool that **magnifies** distant objects so that they appear larger and closer.

Galileo Galilei (1564–1642) was a scientist who lived in Italy. In 1609, he improved the telescope and used it to observe the night sky. He could see many more stars through the telescope than with his unaided eyes. He could see mountains and **craters** on the Moon. And he could see that planets were spheres, not just points of light. Then Galileo turned his telescope toward Jupiter. He became the first person to observe moons orbiting another planet.

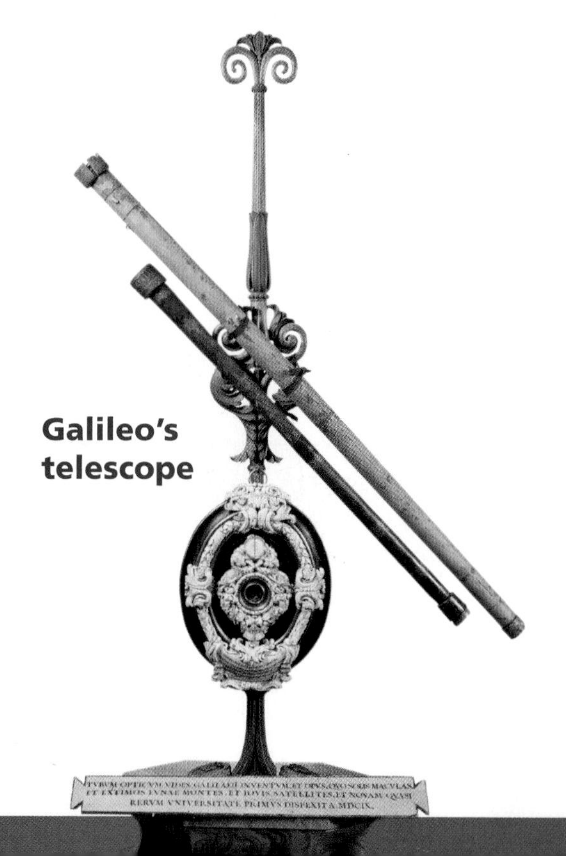
Galileo's telescope

You can see more stars through a telescope than with your unaided eyes.

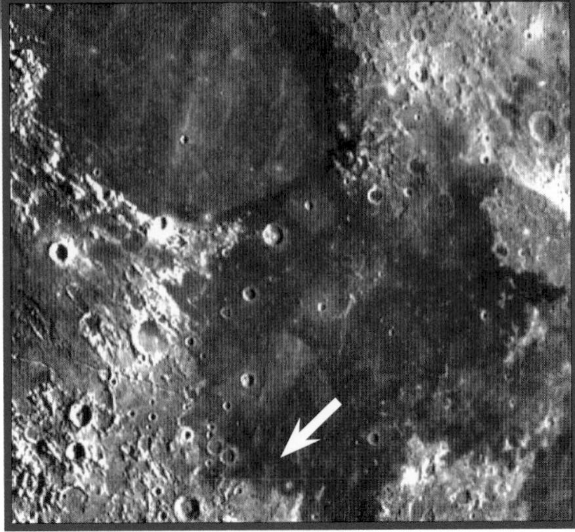

The Apollo 11 landing site on the Moon

As telescopes got more powerful, **astronomers** could see more details on planets. They could also see more stars in the night sky. By the mid-1900s, the surface of the Moon could be studied in detail with telescopes on Earth. Scientists used pictures taken through telescopes to plan the first Moon landing in 1969.

Modern Telescopes

Most telescopes are built on mountain peaks. The telescopes are above most of the dust and pollution in the **air**. And they are far away from city lights. The telescopes are protected inside dome-shaped buildings called **observatories**.

Keck Observatory is on top of Mauna Kea, a 4,205-meter peak on the island of Hawaii.

The space shuttle placed a very important telescope in Earth's orbit in 1990. It is called the Hubble Space Telescope. The Hubble Space Telescope takes pictures of planets and other objects in the **solar system**. It also takes pictures of objects beyond the solar system. Because the telescope orbits above Earth's **atmosphere**, it gets a clear view of outer space.

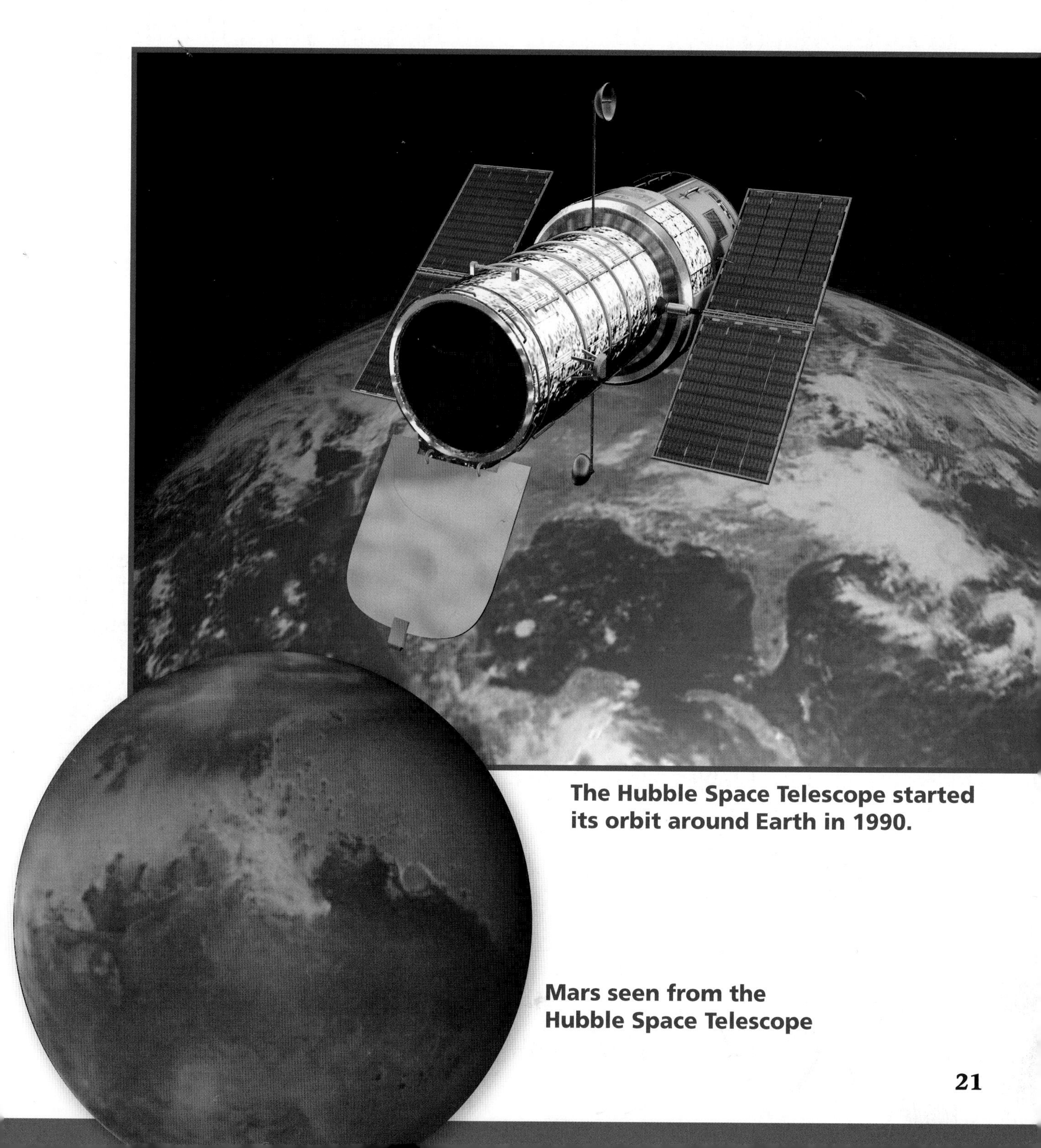

The Hubble Space Telescope started its orbit around Earth in 1990.

Mars seen from the Hubble Space Telescope

When you look up at the sky on a clear, moonless night, you can see about 2,000 stars. But that view changes a lot when you look through the Hubble Space Telescope. You can see millions of stars that are too dim to see with your unaided eyes. Telescopes make distant objects look bigger and closer. With telescopes, astronomers can explore space without leaving Earth.

Part of the Milky Way seen with unaided eyes on a clear night

Part of the Milky Way seen through the Hubble Space Telescope

On April 24, 2015, Hubble was 25 years old. For 25 years, it had provided information about the structure and organization of the universe. Hubble looked into the most distant parts of our **galaxy** to observe the birth of stars.

The Westerlund 2 cluster, a collection of about 3,000 newly-formed stars

Hubble has also allowed astronomers to determine the size and age of the universe. By looking between the stars in the night sky, Hubble observed objects in deep, deep space. Hubble observed that in every direction in the sky, the deep field of view is filled with millions of other galaxies. Each galaxy is made of up billions of stars. Measurements made by Hubble show that the galaxies are all moving away from one another. This is evidence that the universe is expanding at a rapid rate. These observations allowed astronomers to figure out that the universe is between 13 and 14 billion years old.

Each point of light in this image is a galaxy composed of billions of stars.

Thinking about Telescopes

1. Who was Galileo, and what was he the first to do?
2. Why is a telescope a useful tool to an astronomer?
3. Why are modern telescopes built on mountaintops or put into space?

Comparing the Size of Earth and the Moon

Apollo 11 Space Mission

On July 16, 1969, the world's most powerful booster rocket thundered off launch pad 39A at Cape Canaveral, Florida. Perched on top of the mighty Saturn 5 rocket was a tiny command module and a smaller, spindle-legged, lunar module. The mission was Apollo 11. On board were three American astronauts. Neil Armstrong was the mission commander. Michael Collins was the command module pilot. Edwin "Buzz" Aldrin Jr. was the lunar module pilot.

The goal of the Apollo 11 mission was to land two men on the Moon and return them safely to Earth. The mission was complex. It involved the development of many new technologies, including some of the most advanced engineering ever attempted by humans. The 36-story-tall Saturn 5 three-stage rocket was the largest, most powerful booster rocket ever designed.

The first stage of the rocket lifted the 3,000-ton spacecraft off Earth's surface. After 8 minutes, the first stage was used up and fell away. At that point, the second stage fired up to propel the spacecraft into orbit 189 kilometers (km) above Earth's surface. After orbiting Earth one and a half times, the third stage of the booster rocket fired up and sent the spacecraft on its way toward the Moon.

Neil Armstrong, Michael Collins, and Buzz Aldrin

The lunar module after it is separated from the command module. The two parts are shown in orbit around the Moon.

The lunar module as it approaches the command module for docking and the return trip to Earth. Earth is seen in earthrise.

As soon as the spacecraft was up to speed, it separated from the third rocket stage and coasted its way to the Moon. Four days later, the spacecraft arrived and moved into orbit around the Moon.

The spacecraft had two separate parts. The first part was the lunar module, the craft that would land on the Moon and later take off from the Moon. The second part was the command module, the craft that would orbit the Moon while waiting for the lunar module to return. The two parts would undock, or separate, during an orbit around the Moon.

When all was ready, Armstrong and Aldrin moved into the lunar module. Mission Control in Houston, Texas, gave the command to the lunar module to start its descent toward the Moon's surface. The two modules separated. *Eagle*, the lunar module, started its long process of slowing down and descending to the Moon's surface. *Columbia*, the command module, stayed in its lunar orbit to await the return of *Eagle* after it completed its mission to the surface.

The preprogrammed descent brought *Eagle* close to the Moon's surface. As *Eagle* approached the landing site, Armstrong and Aldrin could see that they were headed for a pile of boulders. At the last minute, Armstrong took the controls to pilot *Eagle* to a safer landing spot.

Footprints left by the astronauts on the Moon are permanent.

Armstrong took this picture of Aldrin. What can you see in the visor of Aldrin's helmet?

After a few tense seconds, Armstrong guided *Eagle* to a soft, safe landing on the southwestern edge of the Sea of Tranquility. Soon after, Armstrong and Aldrin reported to Mission Control in Texas: "Houston, Tranquility Base here. The *Eagle* has landed!" Dozens of technicians at Mission Control cheered for this amazing event. Humans had arrived safely on the surface of the Moon.

After checking all systems in the lunar module to make sure it was secure and undamaged, Armstrong and Aldrin dressed for a trip outside. The Moon's surface, with no atmosphere, is a deadly place for a person without proper protection. The **temperature** is more than 115 degrees Celsius (°C) in the sunshine and –173°C in the shade. The pressure is 0, and there is no air.

Dressing involved putting on a pressurized space suit that was temperature controlled. The suit provided air and communication. The helmet had a gold-covered lens that could be lowered to protect the astronauts' eyes from dangerous ultraviolet rays from the Sun.

At 10:39 p.m. eastern daylight time, Armstrong squeezed out of the exit hatch onto the ladder leading down to the Moon's surface. As he hopped from the lowest rung onto the Moon's surface, he said, "That's one small step for a man, one giant leap for mankind."

The lunar soil onto which Armstrong stepped was like powder. The bulky, stiff suit worked perfectly. Armstrong was comfortable and able to move around easily. Aldrin joined him on the Moon's surface, and together they began their tasks. They set up several experiments on the surface. They put up an American flag, took photos of the terrain, and collected samples of lunar rocks and soil.

After 2 hours and 21 minutes of exploring the Moon's surface, the astronauts gathered their equipment and scientific samples, including 108 kilograms (kg) of Moon rocks, and returned to the lunar module. They repressurized the cabin and settled in for some much needed rest before leaving the Moon's surface.

Aldrin climbs down to step on the Moon's surface.

Armstrong and Aldrin left the American flag on the Moon.

After 7 hours of rest, Mission Control sent the astronauts a wake-up call. Two hours later, they fired the ascent rocket that propelled the lunar module upward. *Eagle* reunited with *Columbia,* which had been orbiting while *Eagle* was on the Moon's surface.

Once the two spacecrafts were reunited, the landing crew transferred to *Columbia*. No longer needed, *Eagle* was left behind in lunar orbit and probably crashed into the Moon in the next few months. Then *Columbia* used its rockets to start its voyage back to Earth. After the long ride home, *Columbia* moved into Earth's orbit. When the time and location were right, rockets fired to push *Columbia* out of orbit and into Earth's atmosphere.

Soon after, huge parachutes opened to slow *Columbia*'s reentry. The historic mission came to a successful end on July 24, when *Columbia* splashed down safely in the Pacific Ocean. They landed only 24 km from the recovery ship waiting for their return.

Six more Apollo missions followed this adventure. The last mission, Apollo 17, was in December 1972. A total of 12 people have walked on the Moon. The Moon is the only **extraterrestrial** object that humans have visited.

***Columbia* safely landed in the Pacific Ocean.**

How Did Earth's Moon Form?

Counting out from the Sun, Earth is the first planet with a **satellite**, or moon. Mercury and Venus, closer to the Sun, don't have moons. Mars, the fourth planet from the Sun, has two moons. Earth probably didn't have a moon at first. Earth got its Moon early in Earth's history as a result of a gigantic collision. The event might have happened about 4.5 billion years ago. This is how it might have happened.

Earth formed from **gas** and dust in the solar system. **Gravity** pulled the gas and dust together to form the planet. As soon as it formed, Earth traveled around the Sun in an almost-circular orbit.

The early solar system was messy. It had lots of large rocks and debris flying around in it. Some of the rocks, called planetesimals, were huge. Scientists now think that one of these planetesimals, the size of Mars, started heading for Earth.

Imagine you had been on Earth to witness the event. The planetismal first appeared as a dot in the sky. Over a period of days and weeks, it appeared bigger and bigger. Then it completely blocked the view from Earth in that direction. Finally, it struck, traveling at perhaps 40,000 kilometers (km) per hour. The collision lasted several minutes.

The crash created a chain of events. First, the impact seemed to destroy the incoming object. The planetesimal turned to gas, dust, and a few large chunks of rock. Some surviving chunks traveled deep into the interior of Earth. A large portion of Earth was destroyed as well. The energy that resulted from the crash produced a huge explosion. Earth itself might have been in danger of being blasted apart.

Second, the explosion threw a tremendous amount of **matter** into motion. Some of this debris flew far out into space. Other matter flew up into the air and then returned to Earth. This debris came in many sizes. Some of it was huge rocks that immediately returned to Earth. A short time later, smaller granules of different sizes fell to Earth. Months or even years later, some of the debris was still floating up in the air.

A large portion of the debris didn't fly off into space, and it didn't return to Earth. It began orbiting Earth. This orbiting debris formed a disk, like the rings of Saturn. The ring was probably about two Earth **diameters** from the surface of Earth. Right away, the pieces of debris started to attract one another. Gradually, they formed into larger and larger chunks. After several weeks, the chunks of debris formed Earth's Moon.

Earth now had a satellite where previously there was none. It must have been quite a sight up there only about 30,000 km above Earth. Today the Moon is much farther above Earth, about 385,000 km.

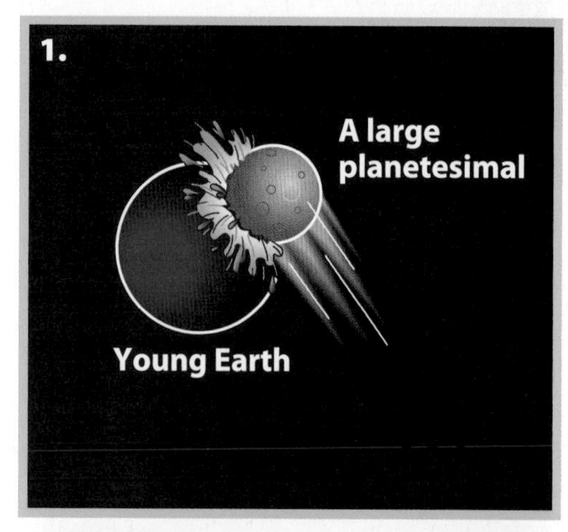

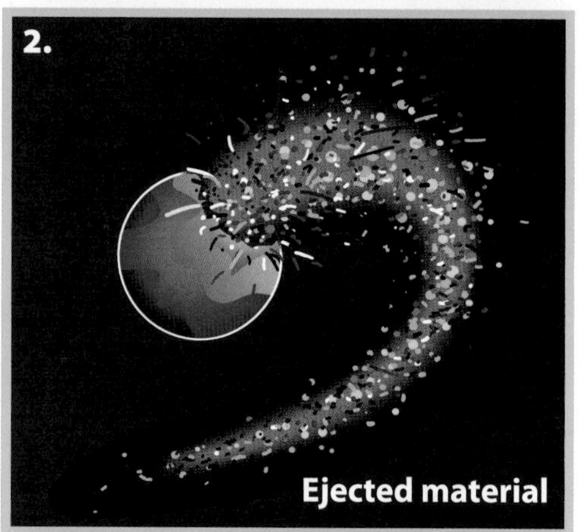

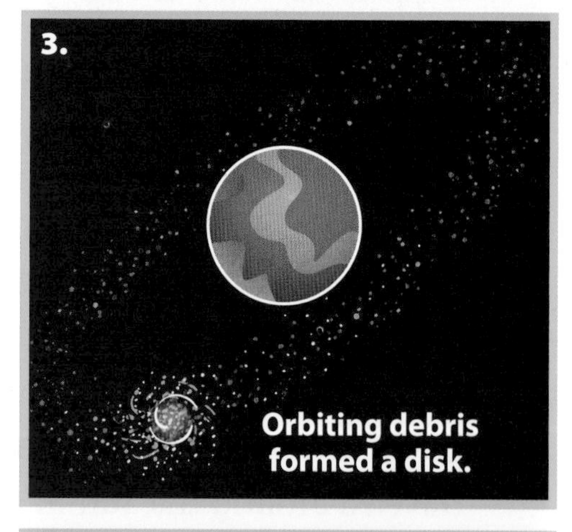

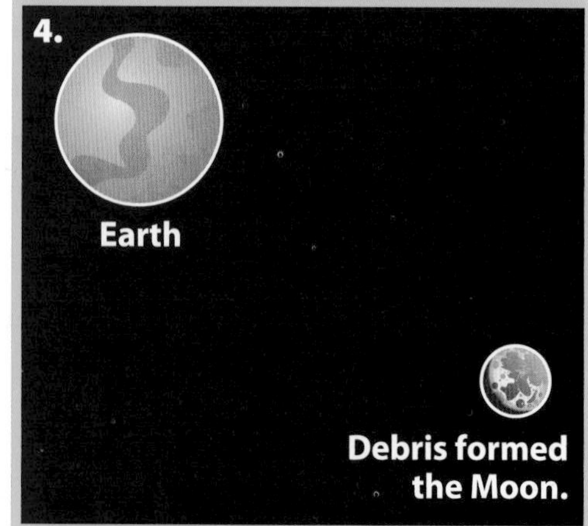

A representation of how the Moon might have formed

Changing Moon

Earth has one large satellite. It is called the Moon. The Moon completes one orbit around Earth every 28 days. One complete orbit is also called a **cycle**.

The Moon is the second-brightest object in the sky. It shines so brightly that you can see it even during the day. But did you know that the Moon doesn't make its own light? The light you see coming from the Moon is **reflected** sunlight. Sunlight reflected from the Moon is what we call moonlight.

The Moon is a sphere. When light shines on a sphere, the sphere is half lit and half dark. Wherever you position the sphere, if the light source is shining on it, one half will always be lit and the other half dark.

The same is true for the Moon. It is always half lit and half dark. The half that is lit is the side toward the Sun. The half that is dark is the side away from the Sun.

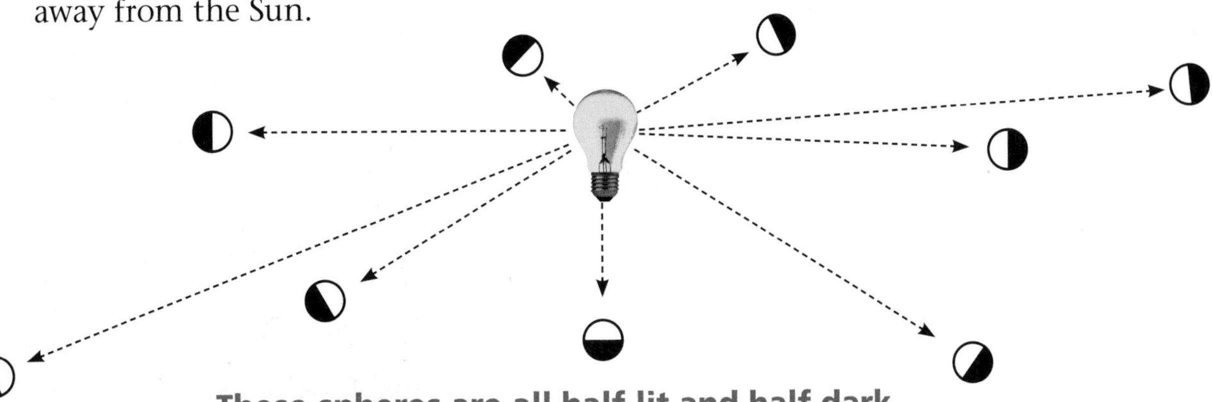

These spheres are all half lit and half dark.

The Moon's Position

From Earth, the Moon never looks the same 2 days in a row. Its appearance changes all the time. Sometimes it looks like a thin sliver, and sometimes it looks completely round. Why does the Moon's appearance change?

The Sun is in the center of the solar system. The planets orbit the Sun. The Moon orbits Earth.

It takes 4 weeks for the Moon to orbit Earth. Where is the Moon during those 4 weeks? Let's take a look from out in space.

We'll start the observations when the Moon is at position 1 between Earth and the Sun.

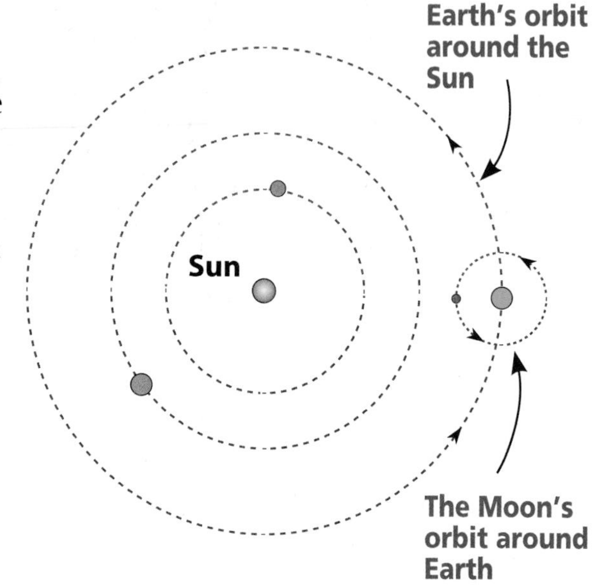

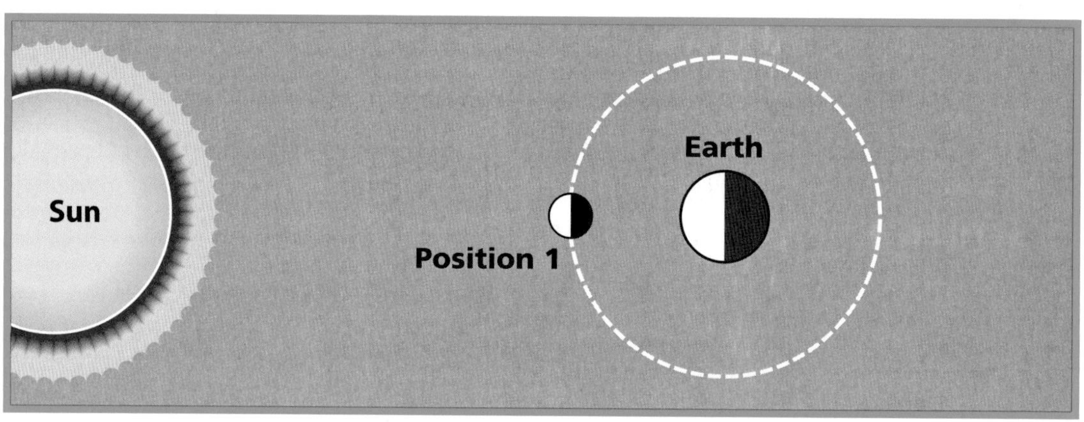

The Moon orbits Earth in a counterclockwise direction. After 1 week, the Moon has moved to position 2.

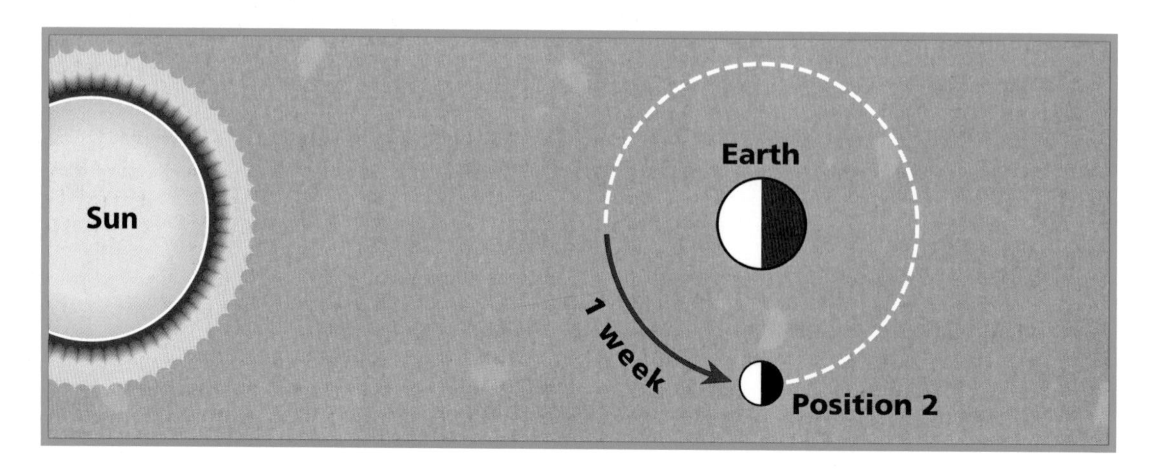

After 2 weeks, the Moon has moved to position 3, on the other side of Earth. The Moon has traveled halfway around Earth.

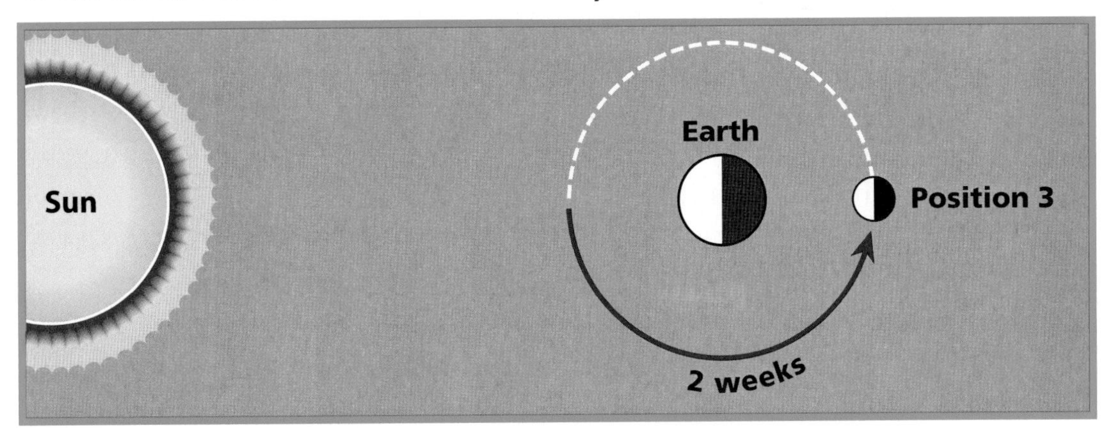

After 3 weeks, the Moon has moved to position 4. It is now three-quarters of the way around Earth.

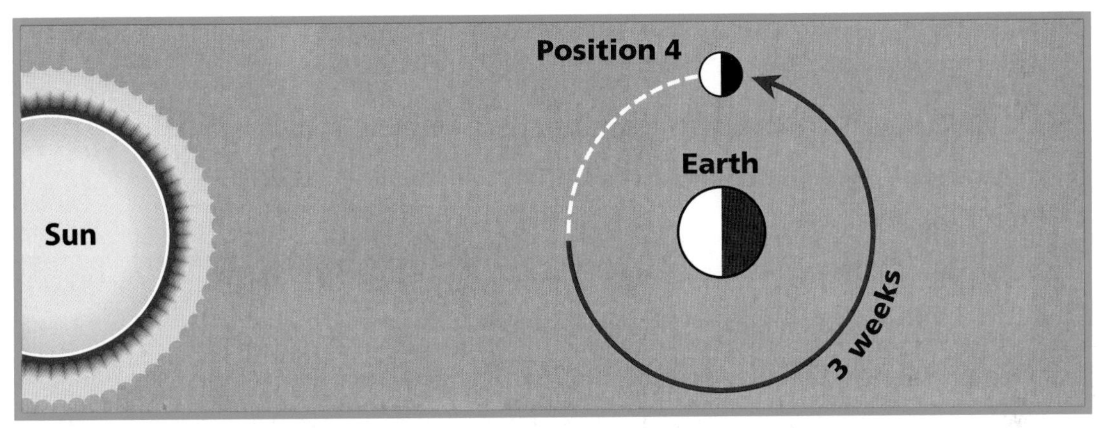

In another week (a total of 4 weeks), the Moon has returned to position 1. It has completed one **lunar cycle**.

Look at the Moon in each of the illustrations. You will see that the lit side is always toward the Sun. In each position during the lunar cycle, the Moon's bright side is always toward the Sun.

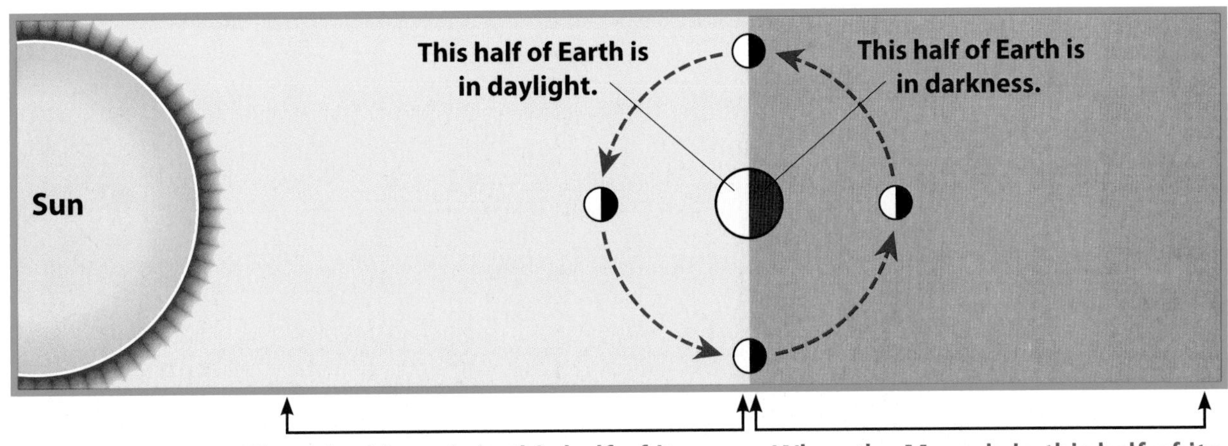

When the Moon is in this half of its orbit, we see the Moon during the day.

When the Moon is in this half of its orbit, we see the Moon during the night.

The Moon's Appearance

The shape of the Moon doesn't change. It is always a sphere. The amount of the Moon that is brightly lit doesn't change. Half of the Moon is always lit by the Sun. What changes is how much of the lit half is visible from Earth. You might see just a tiny bit of the lit half. Or you might see the entire lit half. The lit portion you see from Earth changes in a predictable way. The different shapes you see have been named, and each one is called a **phase**.

Let's look at the phase of the Moon in position 1. The Moon is between the Sun and Earth. When you look up at the Moon from Earth, what do you see? Nothing. All of the lit half of the Moon is on the other side. This is the **new Moon**. The new Moon has no light visible from Earth. The new Moon is shown as a black circle.

Let's move forward 2 weeks. The Moon has continued in its orbit and is in position 3. What do you see when you look up at the Moon? The whole lit side of the Moon is visible from Earth. This is the full Moon. The full Moon is shown in the illustration as a white circle.

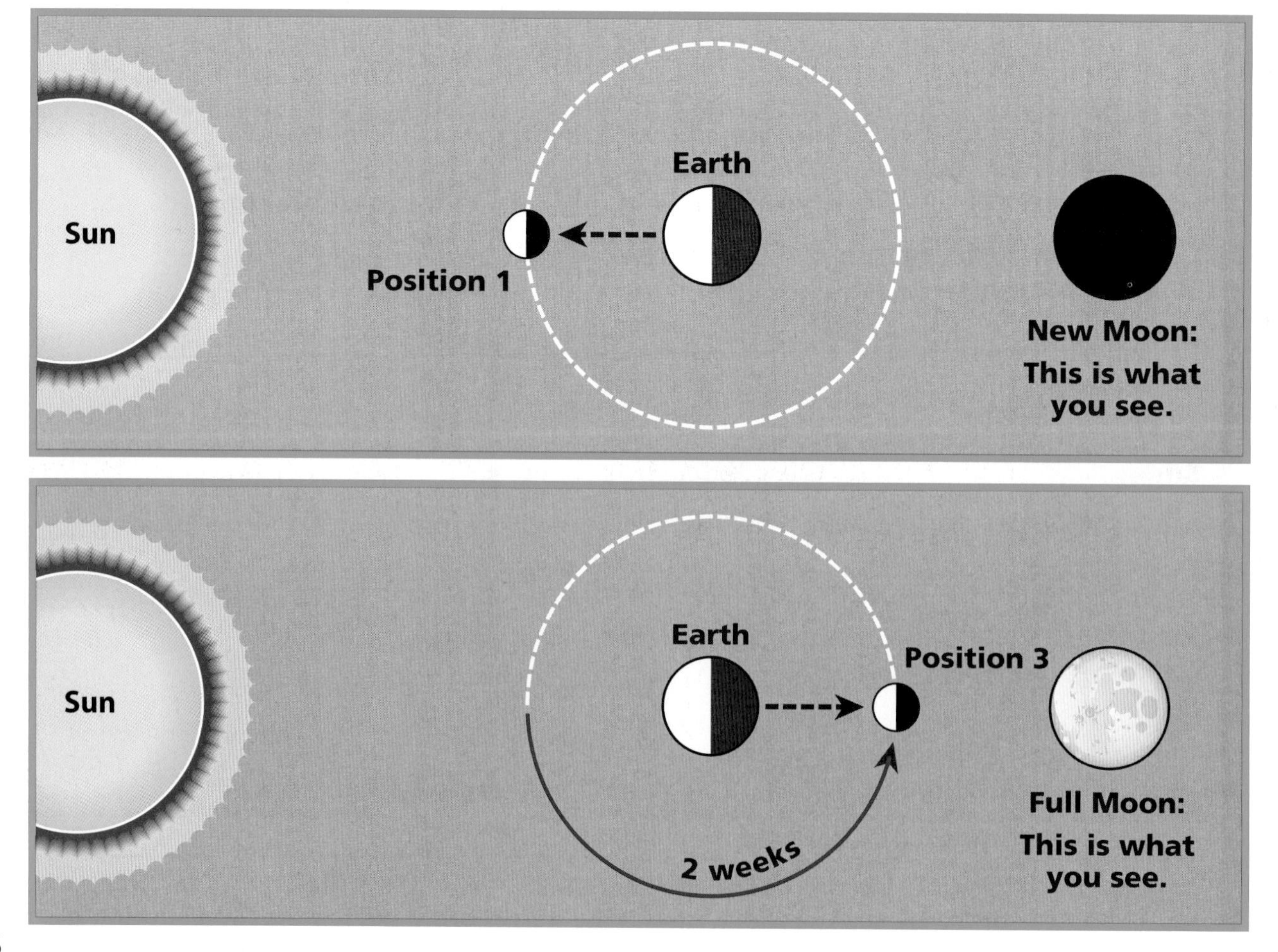

Now let's look at positions 2 and 4. At both positions, you see half of the lit part of the Moon and half of the dark part of the Moon. At position 2, when you look up at the Moon, the lit part is on the right side. At position 4, the lit part of the Moon is on the left side. Position 2 is the **first-quarter Moon**. Position 4 is the **third-quarter Moon**.

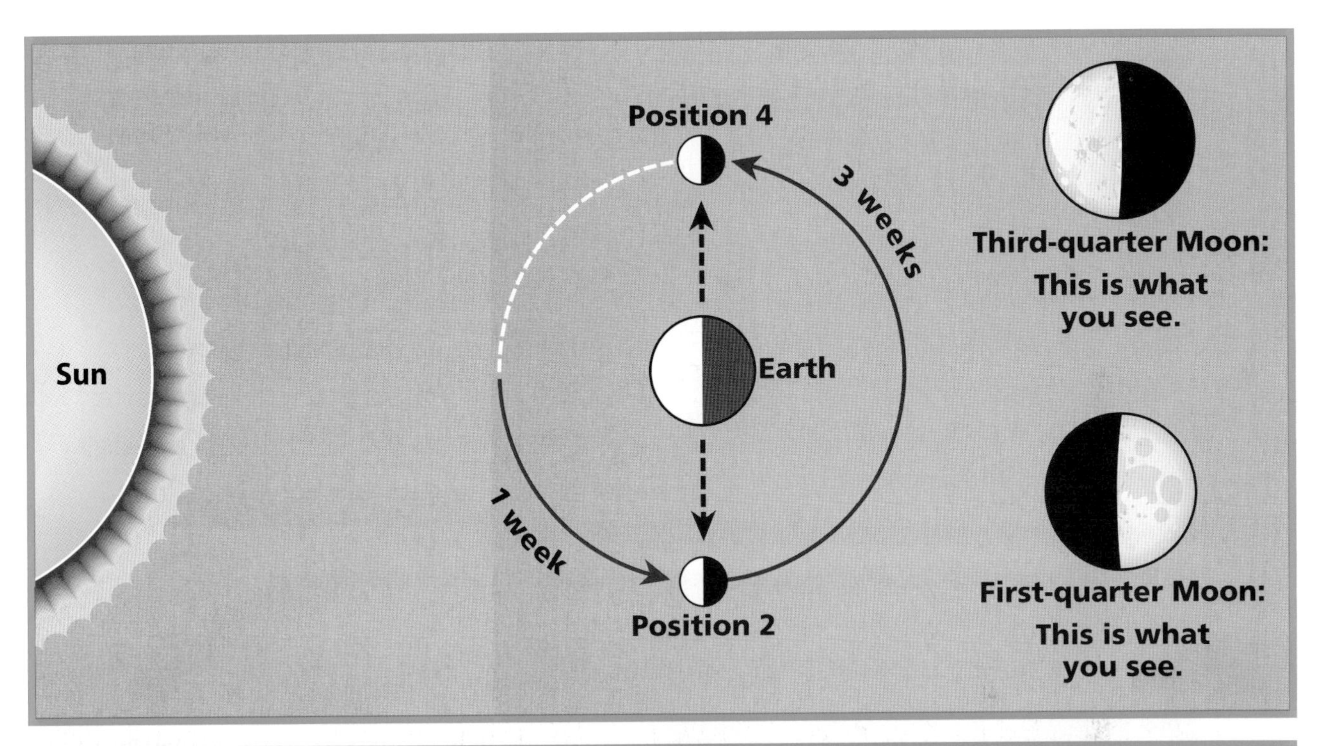

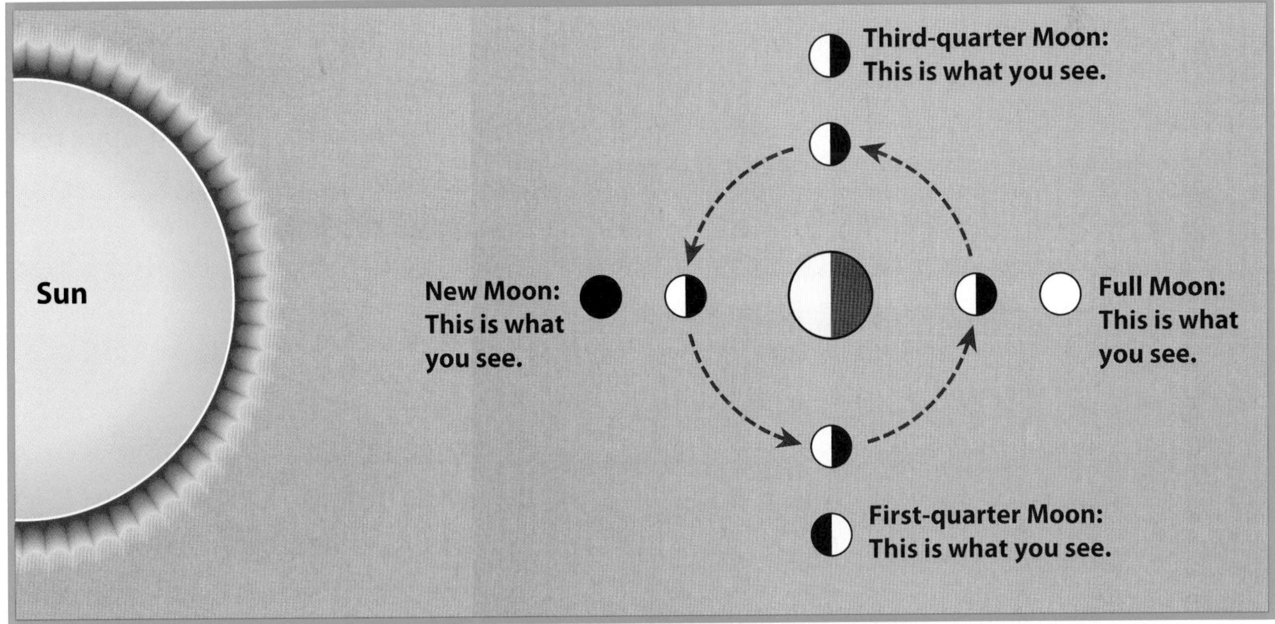

This is the Moon's orbit during one lunar cycle.

Lunar Cycle

The new Moon is invisible for two reasons. First, no light is coming to your eyes from the Moon. The lit side is facing away from Earth. Second, to look for the new Moon, you would have to look toward the Sun. The glare is too bright to see the Moon. (And remember, you should never look directly at the Sun.)

Day 0: New Moon

Day 3: Waxing crescent Moon

Three days later, the Moon has moved in its orbit, and it is visible. The first sighting of the Moon after a new Moon is a tiny sliver of visible light. The curved shape is called the crescent Moon.

Day 5: Waxing crescent Moon

On day 5, the Moon looks larger. About one-quarter of the Moon is now bright. Each day the visible bright part of the Moon is a little larger. We say the Moon is **waxing** when it appears to be growing.

Day 6: Waxing crescent Moon

By day 6, almost half of the Moon appears brightly lit. This is the last day of the waxing crescent Moon. Tomorrow the Moon will appear as the first-quarter Moon.

The first-quarter Moon is the phase seen on day 7. The Moon has completed the first quarter of its lunar cycle. Observers on Earth see half the sunlit side of the Moon and half the dark side of the Moon. The brightly lit side is on the right side.

Day 7: First-quarter Moon

On day 9, you can see more than half the sunlit side of the Moon. The Moon appears to be oval shaped. A Moon phase that is larger than a quarter but not yet full is called a **gibbous Moon**. Because the Moon is still getting bigger, it is a waxing gibbous Moon.

Day 9: Waxing gibbous Moon

On day 11, the Moon is almost round. It is still a waxing gibbous Moon. Observers on Earth can see most of the sunlit half of the Moon. They can see only a small sliver of the dark side of the Moon. Can you see the dark crescent?

Day 11: Waxing gibbous Moon

On day 14, you can see the whole sunlit side of the Moon. This is the full-Moon phase. A full Moon always rises at the same time the Sun sets.

Day 14: Full Moon

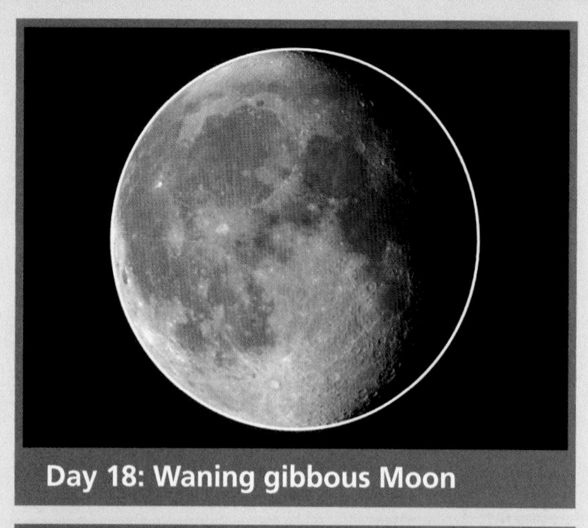
Day 18: Waning gibbous Moon

Each day after the full Moon, the bright part of the Moon gets smaller. Getting smaller is called **waning**. On day 18, the Moon looks oval again. Because it is still between full-Moon phase and quarter phase, it is still a gibbous Moon, a waning gibbous Moon.

Day 21: Third-quarter Moon

On day 21, the Moon has completed three-quarters of its orbit around Earth. The Moon appears as the third quarter, again half bright and half dark. But notice that the bright side of the third-quarter phase is on the left. Compare the appearance of the third-quarter Moon and the first-quarter Moon.

Day 24: Waning crescent Moon

As the Moon starts the last 7 days of its orbit, it returns to crescent phase. But because it is getting smaller each day, it is the waning crescent phase. By day 24, an observer on Earth sees just a small part of the sunlit side of the Moon. A lot of the dark side is visible again.

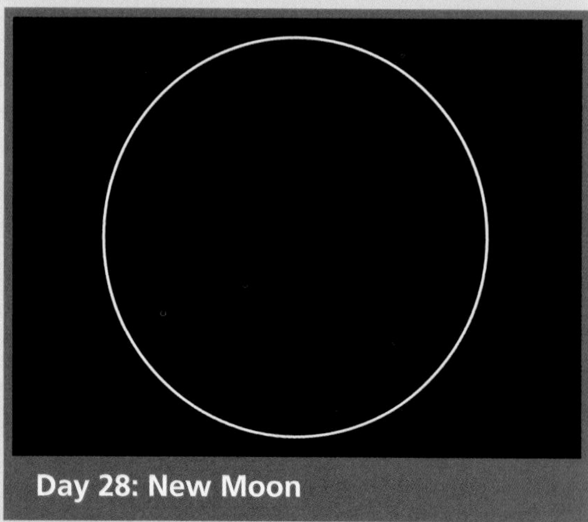
Day 28: New Moon

On about day 28, the Moon has completed one lunar cycle. It is back at its starting point. It is at the new-Moon phase again. The night sky is moonless. The day sky has no Moon. For a couple of days, observers on Earth can't see the Moon.

Then, in the evening sky, just after sunset, the Moon reappears. It is a thin, silver-colored crescent. And if you are in the right place at the right time, you could see something special. It is a bright crescent on the edge of a dim full Moon. It is called the old Moon in the new Moon's arms.

The old Moon in the new Moon's arms

How can you see a bright crescent Moon and a pale full Moon at the same time? When the Moon appears as a thin crescent, it is almost between Earth and the Sun. A lot of light reflects from Earth onto the Moon. The whole Moon is dimly lit by earthshine.

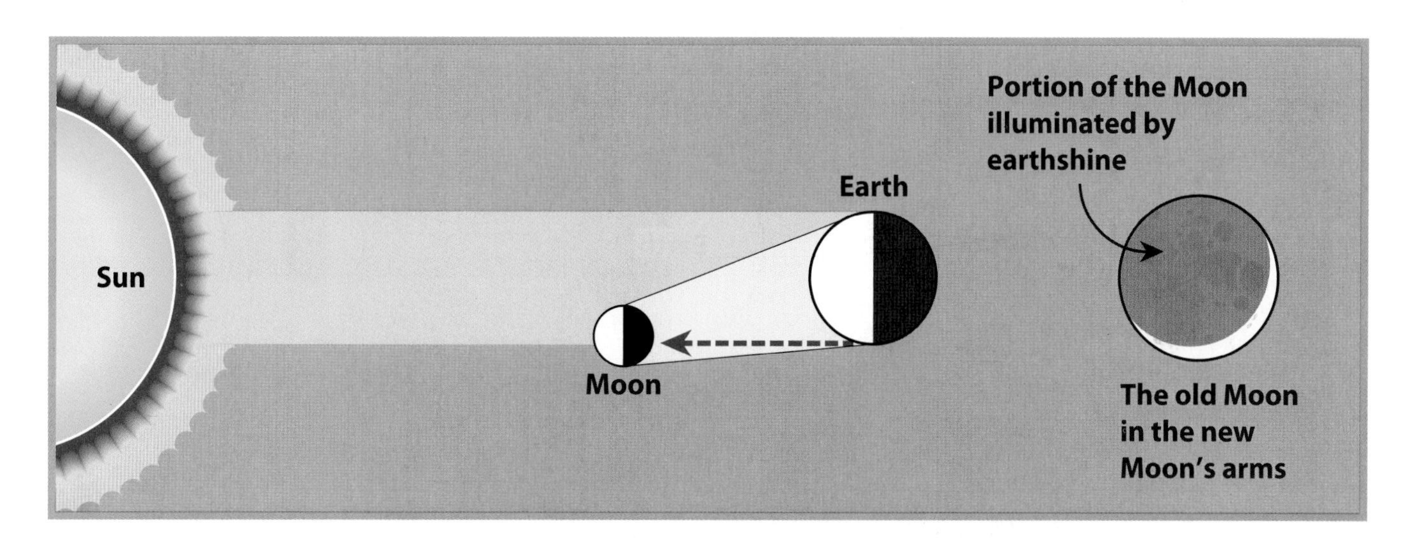

Thinking about the Phases of the Moon

1. How long does it take the Moon to complete one lunar cycle?
2. What is a new Moon, and what causes it?
3. What is the difference between a waxing Moon and a waning Moon?
4. What is the difference between a crescent Moon and a gibbous Moon?
5. Describe the Moon's appearance 1 week, 2 weeks, 3 weeks, and 4 weeks after the new Moon.

Lunar Cycle Diagram

The position of the Moon in its lunar cycle determines the Moon's phase.

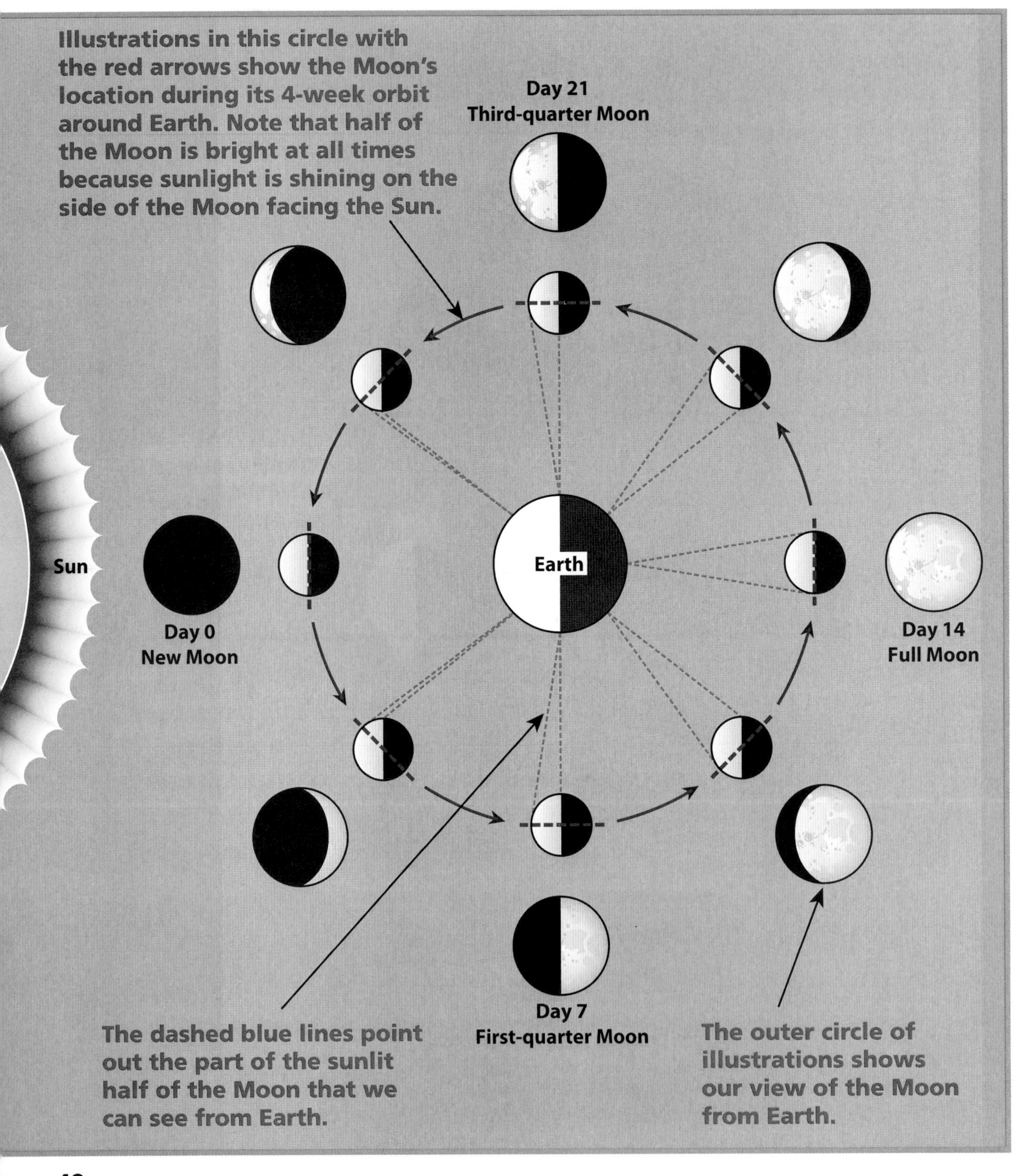

A total solar eclipse

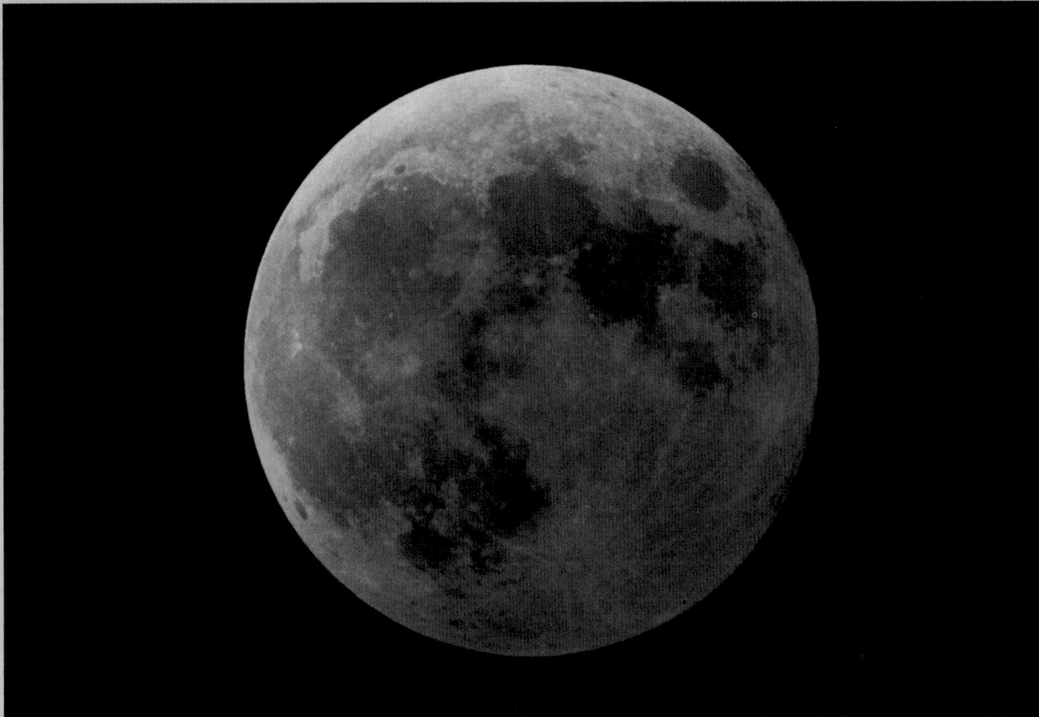

A total lunar eclipse

Eclipses

Occasionally, people on Earth are able to observe a lovely orange-colored eclipse of the Moon (a **lunar eclipse**). Less frequently, they can observe a black-centered eclipse of the Sun (a **solar eclipse**).

What causes these interesting events? When can you see a lunar eclipse? When can you see a solar eclipse?

What Is a Solar Eclipse?

A solar eclipse occurs when the Moon passes exactly between Earth and the Sun. The Moon completely hides the disk of the Sun when this happens. This diagram shows the alignment of Earth, the Moon, and the Sun during a solar eclipse.

A total solar eclipse

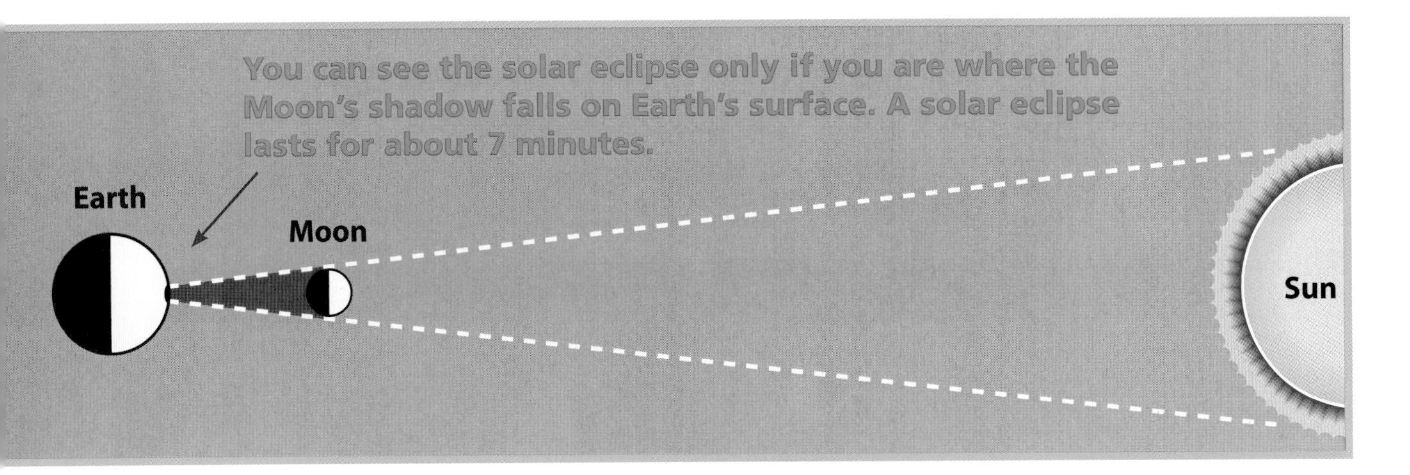

You can see a solar eclipse only on a very small area of Earth's surface. A total eclipse of the Sun is visible for a bit more than 7 minutes, as long as it takes for the disk of the Moon to pass across the disk of the Sun.

The Moon travels around Earth once every month. Why doesn't a solar eclipse occur every month? The Moon's orbit around Earth is not in the same plane as the orbit of Earth going around the Sun. The Moon's orbit is tilted a little bit. Most months, Earth, the Moon, and the Sun are not in a straight line.

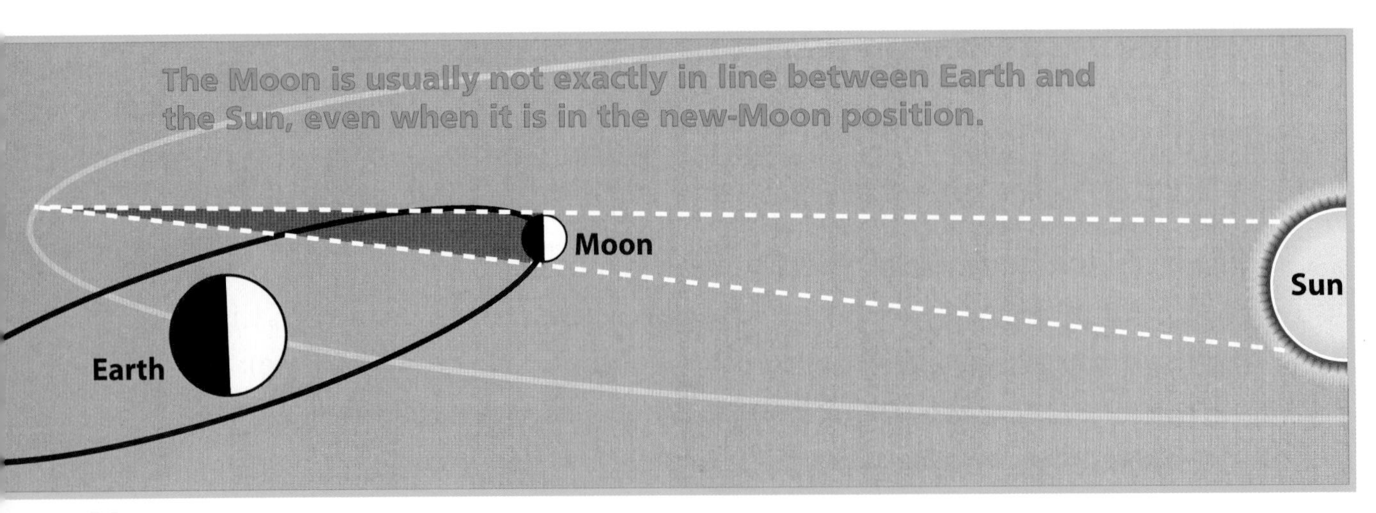

What Is a Lunar Eclipse?

A lunar eclipse occurs when Earth passes exactly between the Moon and the Sun. Earth's shadow completely covers the disk of the Moon when this happens. This diagram shows the alignment of Earth, the Moon, and the Sun during a lunar eclipse.

A lunar eclipse

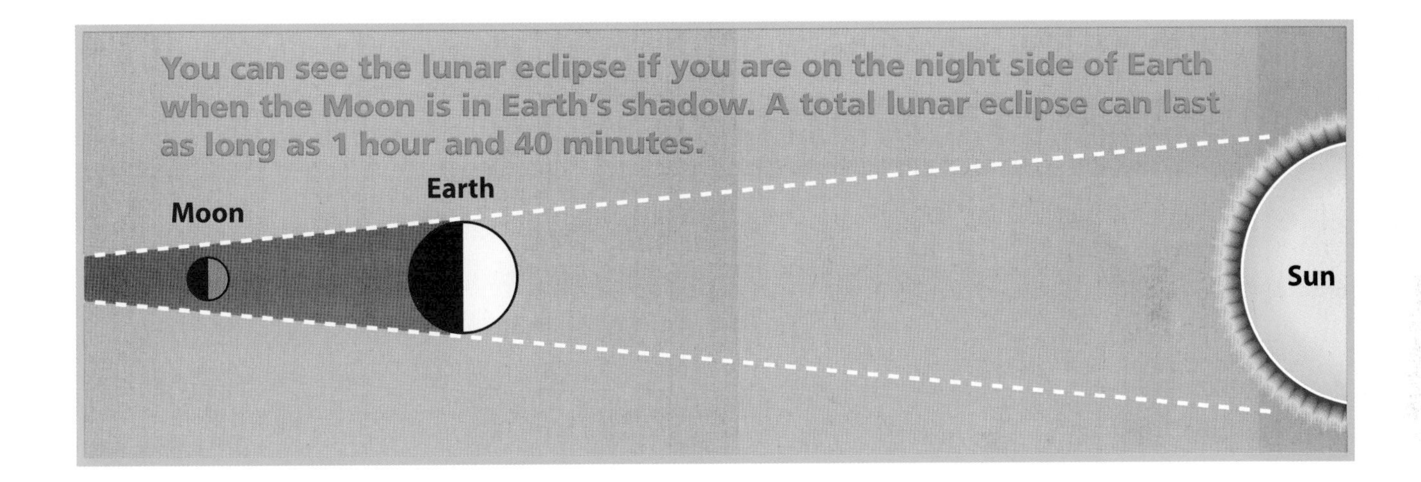

You can see a lunar eclipse from anywhere on Earth where it is night. A total lunar eclipse lasts almost 2 hours and its beautiful red color is safe to view without eye protection.

Why don't we see a lunar eclipse every month? Again, it's because of the tilt of the Moon's orbit around Earth. The Moon's orbit around Earth is not in the same plane as the orbit of Earth going around the Sun. In most months, Earth's shadow does not fall on the Moon.

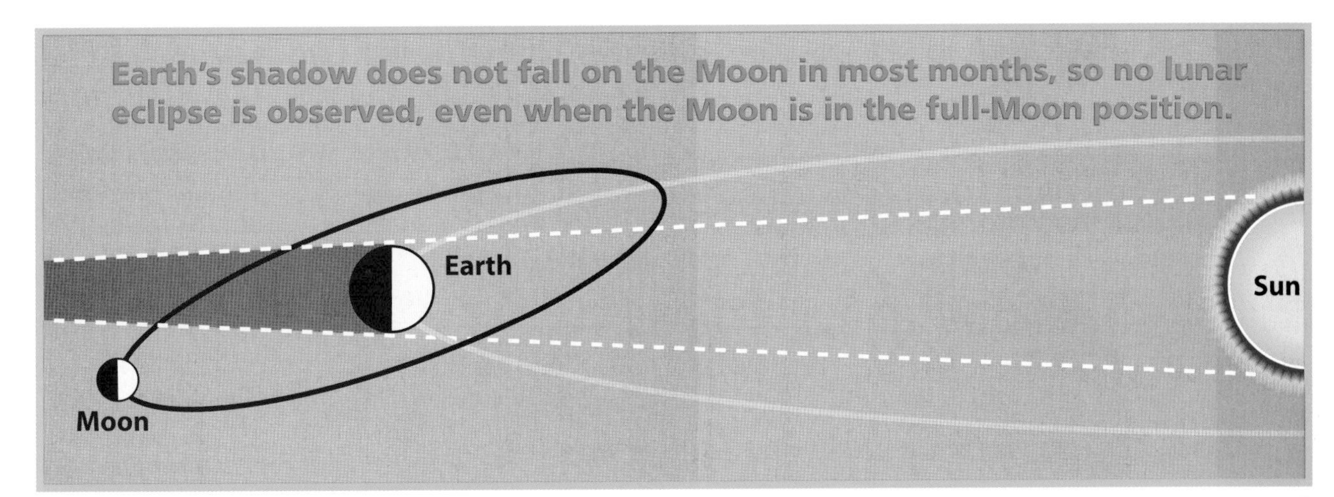

Look at the sequence of photos showing a total lunar eclipse. The Moon moves across Earth's shadow. The diagram below the photos explains what's happening. Note the reddish-brown color of the last photo.

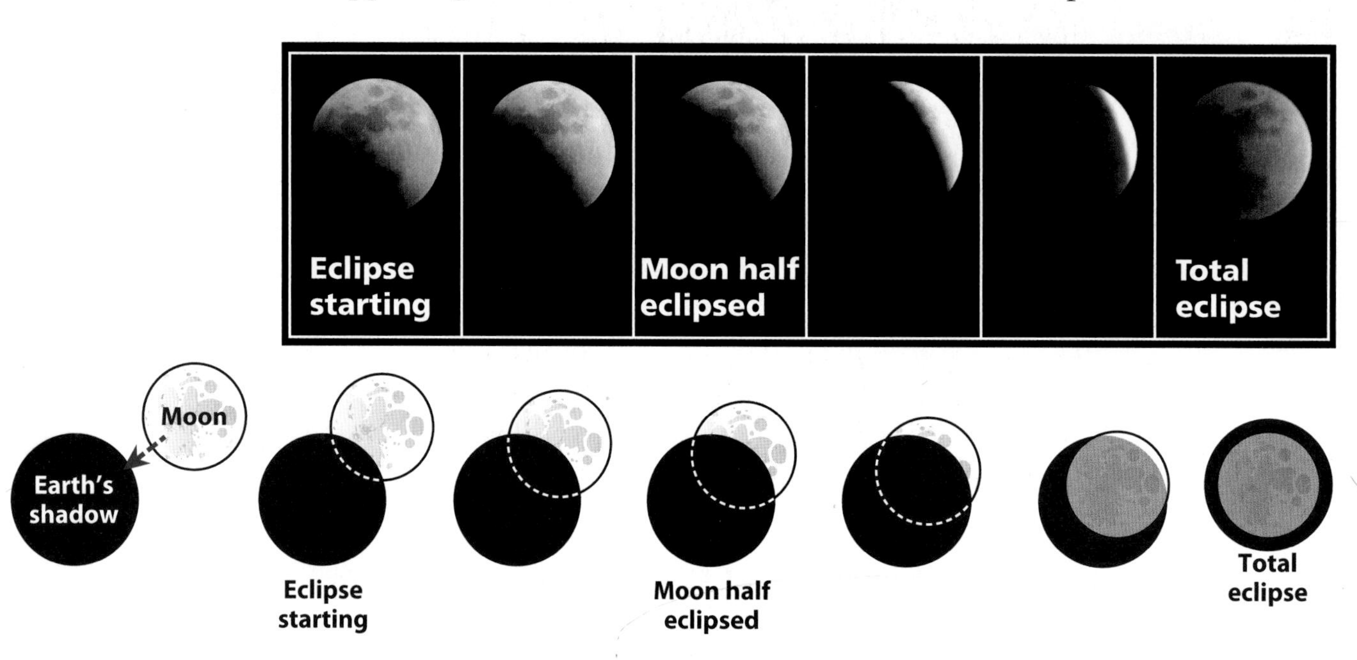

Why is the totally eclipsed Moon reddish-brown? Why is it visible at all? When light passes through Earth's atmosphere, it is bent and scattered by the air. As a result, some reddish light falls on the Moon's surface. This makes the Moon appear reddish-brown. If this bending and scattering did not occur, a totally eclipsed Moon would be invisible because no light would hit the Moon. If no light hit the Moon, no light would be reflected back into our eyes on Earth.

Thinking about Eclipses

1. During what phase of the Moon can you observe a lunar eclipse?
2. During what phase of the Moon can you observe a solar eclipse?

Sizes and distances of solar-system objects are not drawn to scale.

Exploring the Solar System

Imagine you are coming to the solar system as a stranger on a tour. There is a tour guide to provide information. You have a window to look out. The tour is about to start. What will you see?

The first view of the solar system is from space. From here, you can see the whole solar system. The most surprising thing is that the solar system is mostly empty. The matter is concentrated in tiny dots. And the dots are far apart. Most of the dots are planets.

There is a star in the center of the solar system. Four small planets orbit pretty close to the star. These are the rocky **terrestrial planets**.

Next, there is a region of small bits of matter orbiting the star. This is the **asteroid** belt.

Out farther, four large planets orbit the star. These are the **gas giant planets** made of gas.

Beyond the gas giant planets is a huge region of icy chunks of matter called the **Kuiper Belt**. Some of the chunks are big enough to be planets. A dwarf planet, Pluto, is one of the Kuiper Belt objects. Others, called comets, have orbits that send them flying through the rest of the solar system.

The Sun

The Sun is a fairly average star. It is much like millions of other stars in the **Milky Way**. The Sun formed about 5 billion years ago. A cloud of gas began to spin. As it spun, it formed a sphere. The sphere got smaller and smaller. As it got smaller, it got hotter. Eventually, the sphere got so hot that it started to radiate light and heat. A star was born.

The Sun is made mostly of hydrogen (72 percent) and helium (26 percent). It is huge. The diameter is about 1,384,000 kilometers (km). The diameter is the distance from one side of the Sun to the other through the center. That's about 109 times the diameter of Earth.

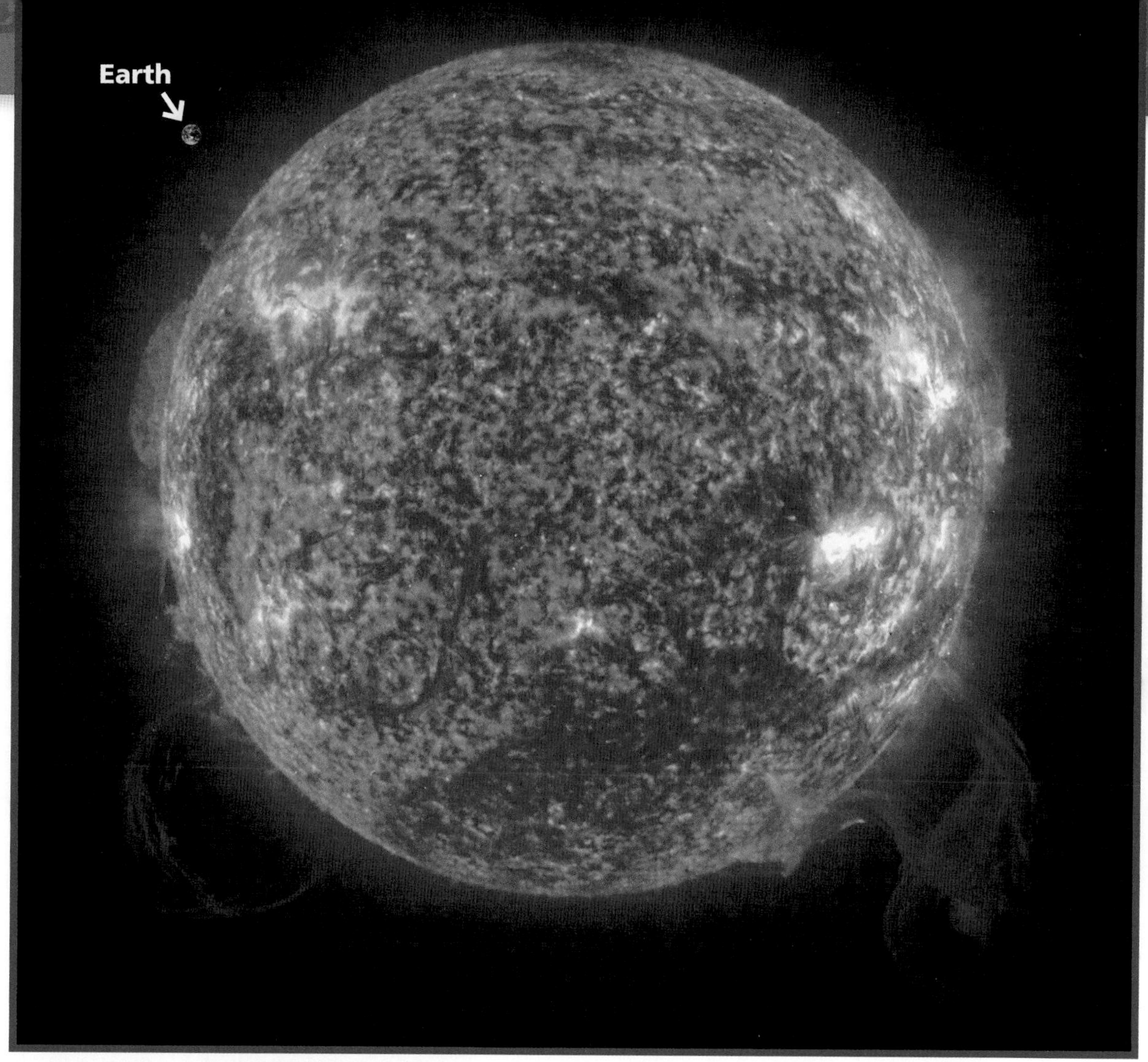

The Sun's diameter is about 109 times the diameter of Earth.

The Sun is incredibly hot. Scientists have figured out that the temperature at the center of the Sun is 15,000,000 degrees Celsius (°C). The temperature of the Sun's surface is lower, about 5,500°C. Hydrogen is constantly changing into helium in **thermonuclear reactions**. These reactions create heat and light. About 3.6 tons of the Sun's **mass** is being changed into heat and light every second. This energy radiates out from the Sun in all directions. A small amount of it falls on Earth.

Another name for the Sun is Sol. That's why the whole system of planets is called the solar system. The solar system is named for the ruling star. The Sun rules because of its size. It has 99.8 percent of the total mass of the solar system. All the other solar-system objects travel around the Sun in predictable, almost-circular orbits. The most obvious objects orbiting the Sun are the planets.

Terrestrial Planets

The terrestrial planets are the four planets closest to the Sun. They are small and rocky.

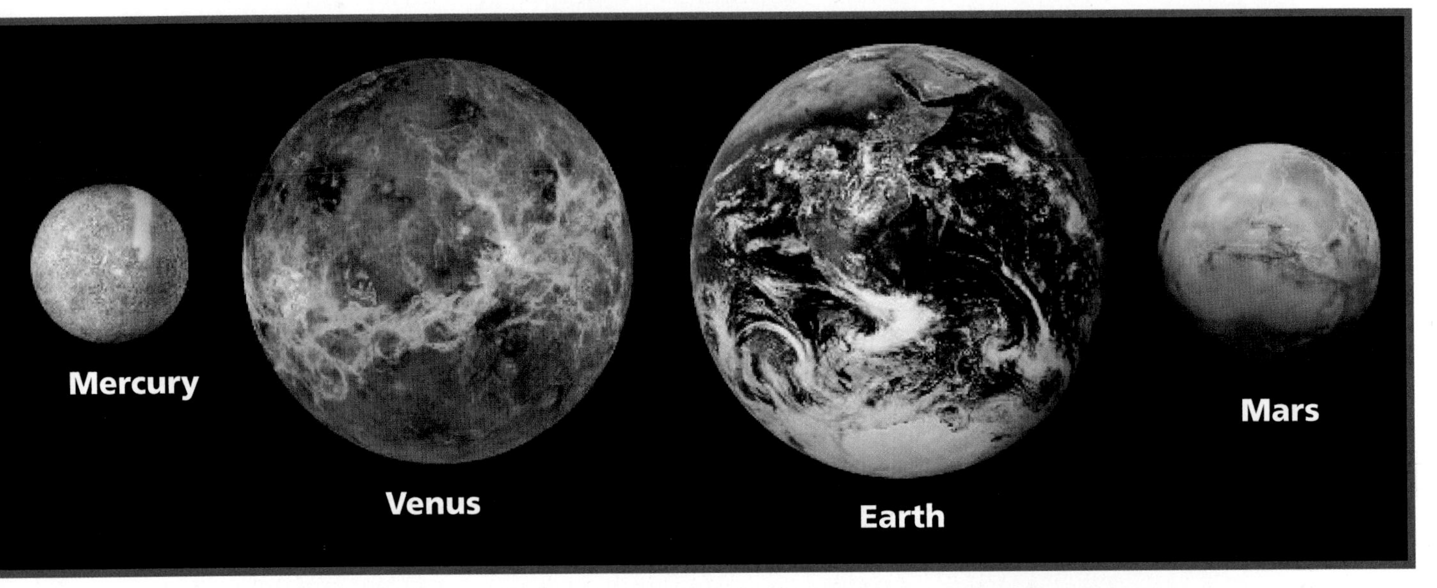

Relative sizes of the terrestrial planets

Mercury

Mercury is the planet closest to the Sun. Mercury is smaller than Earth and has no satellite (moon). By human standards, it is an uninviting place. Mercury is very hot on the side facing the Sun and very cold on the dark side. It has no atmosphere or water.

Mercury is covered with craters. The craters are the result of thousands of collisions with objects flying through space. The surface of Mercury looks a lot like Earth's Moon.

Mercury is the planet closest to the Sun.

Venus

Venus is the second planet from the Sun. Venus is about the same size as Earth and has no satellites. The surface of Venus is very hot all the time. It is hot enough to melt lead, making it one of the hottest places in the solar system.

There is no **liquid** water on Venus. But Venus does have an atmosphere of carbon dioxide. The dense, cloudy atmosphere makes it impossible to see the planet's surface. Modern radar, however, allows scientists to take pictures through the clouds. We now know that the surface of Venus is dry, cracked, and covered with volcanoes.

The surface of Venus is dry and covered with volcanoes.

Earth

Earth is the third planet from the Sun. Earth has a moderate, or mild, temperature all the time. It has an atmosphere of nitrogen and oxygen, and it has liquid water. As far as we know, Earth is the only place in the universe that has life. Earth also has one large satellite called the Moon. The Moon orbits Earth once a month. The Moon is responsible for the tides in Earth's ocean. The Moon is the only extraterrestrial place humans have visited.

Earth is 150 million km from the Sun. This is a huge distance. It's hard to imagine that distance, but think about this. Sit in one end zone of a football field and curl up into a ball. You are the Sun. A friend goes to the other end zone and holds up the eraser from a pencil. That's Earth. Get the idea? Earth is tiny, and it is a long distance from the Sun. Still, the light and heat that reach Earth provide the right amount of energy for life as we know it.

The Moon orbits Earth once a month.

Water frost on the surface of Mars

A robotic lander exploring Mars

Mars

Mars is the fourth planet from the Sun. It has two small satellites, Phobos and Deimos. Mars is a little like Earth, except it is smaller, colder, and drier. There are some places on Mars that are like Death Valley in California. Other places on Mars are more like Antarctica, and others are like the volcanoes of Hawaii.

Mars is sometimes called the red planet because of its red soil. The soil contains iron oxide, or rust. The iron oxide in the soil tells scientists that Mars probably had liquid water at one time. But liquid water has not been on Mars for 3.5 billion years. It has frozen water in polar ice caps that grow and shrink with its seasons.

Mars is likely the next place humans will visit. But exploring Mars will not be easy. Humans can't breathe the thin atmosphere of carbon dioxide. And explorers will need to wear life-support space suits for protection against the cold.

Several robotic landers, including *Viking, Spirit, Opportunity, Sojourner,* and *Curiosity* have observed Mars and sent back information about the surface and presence of water. Evidence suggests that there is a lot of frozen water just under the surface.

Asteroids

Beyond the orbit of Mars are millions of chunks of rock and iron called asteroids. They all orbit the Sun in a region called the asteroid belt. The asteroid belt surrounds the terrestrial planets. The planets farther out are quite different from the terrestrial planets.

Some asteroids even have moons. When the spacecraft *Galileo* flew past asteroid Ida in 1993, scientists were surprised to discover it had a moon. They named it Dactyl. The largest object in the asteroid belt is Ceres, a dwarf planet. It is about 960 km around.

Asteroid Ida with moon Dactyl

Gas Giant Planets

The four planets farthest from the Sun are the gas giant planets. They do not have rocky surfaces like the terrestrial planets. So there is no place to land or walk around on them. They are much bigger than the terrestrial planets. What we have learned about the gas giant planets has come from probes launched on rockets to fly by and orbit around the giants. Even though they are made of gases, each gas giant planet is different.

Relative sizes of the gas giant planets

Jupiter and its four largest moons

Jupiter

Jupiter is the fifth planet from the Sun. It is the largest planet in the solar system. It is 11 times larger in diameter than Earth. Scientists have found 67 moons orbiting Jupiter. The four largest moons are Ganymede, Callisto, Io, and Europa.

Jupiter's atmosphere is cold and poisonous to life. It is mostly hydrogen and helium. Jupiter's stripes and swirls are cold, windy clouds of ammonia and water. Its Great Red Spot is a giant storm as wide as three Earths. This storm has been raging for hundreds of years. On Jupiter, the atmospheric pressure is so strong that it squishes gas into liquid. Jupiter's atmosphere could crush a metal spaceship like a paper cup.

An artist's drawing of Jupiter, its moon Io, and the *Galileo* spacecraft

Saturn

Saturn is the sixth planet from the Sun. It is the second largest planet and is very cold. At least 60 satellites orbit Saturn. Most of the planet is made of hydrogen, helium, and methane. It doesn't have a solid surface. It has clouds and storms like Jupiter, but they are harder to see because they move so fast. Winds in Saturn's upper atmosphere reach 1,825 km per hour.

The most dramatic feature of Saturn is its ring system. The largest ring reaches out 200,000 km from Saturn's surface. The rings are made of billions of small chunks of ice and rock. All the gas giant planets have rings, but they are not as spectacular as Saturn's.

Close-up of the rings of Saturn

Uranus

Uranus is the seventh planet from the Sun. Uranus has 27 moons and 11 rings. Uranus is very cold and windy, and would be poisonous to humans. It is smaller and colder than Saturn.

Uranus has clouds that are extremely cold at the top. Below the cloud tops, there is a layer of extremely hot water, ammonia, and methane. Near its core, Uranus heats up to 4,982°C. Uranus looks blue because of the methane gas in its atmosphere.

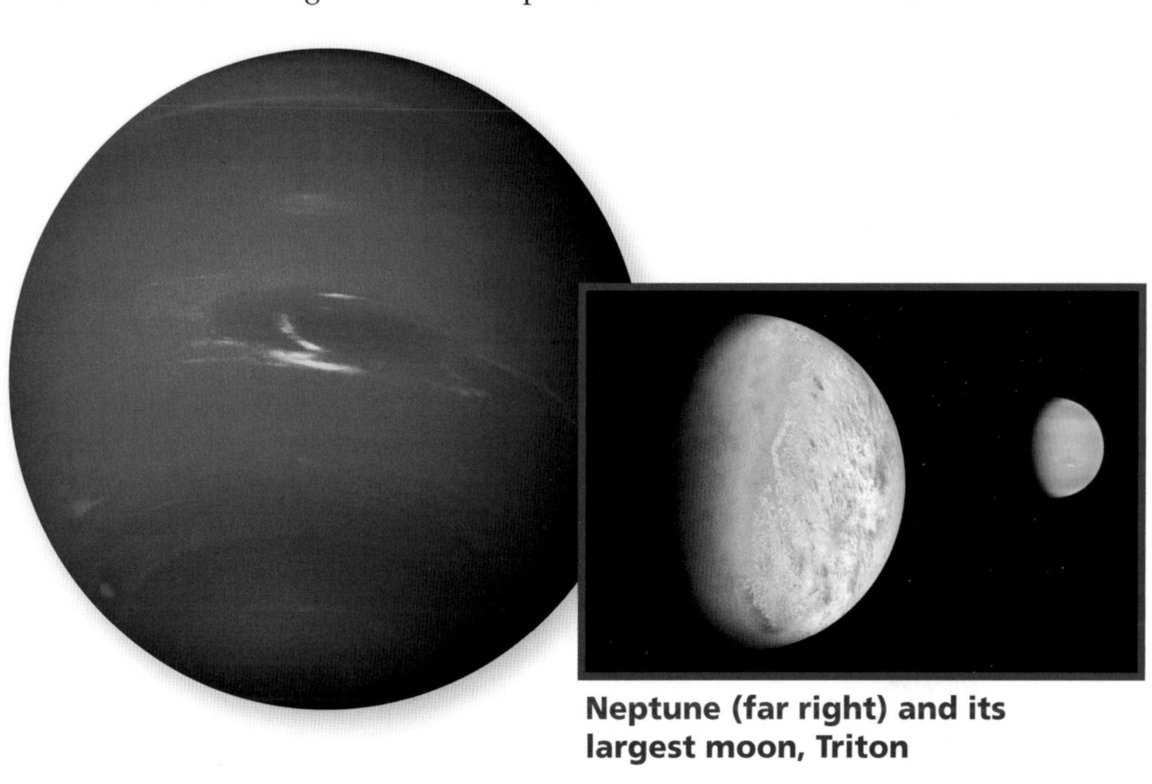

Neptune (far right) and its largest moon, Triton

Neptune

Neptune is the eighth planet from the Sun. Neptune has 13 moons and 4 thin rings. It is the smallest of the gas giant planets, but is still much larger than the terrestrial planets.

Neptune is made mostly of hydrogen and helium with some methane. It may be the windiest planet in the solar system. Winds rip through the clouds at more than 2,000 km per hour. Scientists think there might be an ocean of super-hot water under Neptune's cold clouds. It does not boil away because of the atmospheric pressure.

Pluto and Charon, one of its moons

Kuiper Belt

Out beyond the gas giant planets is a disk-shaped zone of icy objects called the Kuiper Belt. Some of the objects are fairly large.

Pluto

Pluto is a large Kuiper Belt object. Some scientists considered Pluto a planet because it is massive enough to form a sphere. Others did not consider Pluto a planet. To them, Pluto is one of the large pieces of debris in the Kuiper Belt. Scientists have agreed to call Pluto a dwarf planet.

Pluto has a thin atmosphere. When Pluto is farthest from the Sun, the atmosphere gets so cold that it freezes and falls to the surface. Even though Pluto is smaller than Earth's Moon, it has its own moons. Charon is the largest (about half the size of Pluto). Nix and Hydra are much smaller, and in 2011-12, two even smaller moons named Kerberos and Styx were discovered. And there may be more!

Eris

In July 2005, astronomers at the California Institute of Technology announced the discovery of a new planet-like object. It is called Eris. Like Pluto, Eris is a Kuiper Belt object and a dwarf planet. But Eris is more than twice as far away from the Sun as Pluto is. This picture is an artist's idea of what the Sun would look like from a position close to Eris.

The Sun would look like a bright star from Eris.

Comets

Comets are big chunks of ice, rock, and gas. Sometimes comets are compared to dirty snowballs. Scientists think comets might have valuable information about the origins of the solar system.

Comets orbit the Sun in long, oval paths. Most of them travel way beyond the orbit of Pluto. A comet's trip around the Sun can take hundreds or even millions of years, depending on its orbit. A comet's tail shows up as it nears the Sun and begins to warm. The gases and dust that form the comet's tail always point away from the Sun.

A comet's tail always points away from the Sun.

Comet orbits can cross planet orbits. In July 1994, a large comet, named Comet Shoemaker-Levy 9, was on a collision course with Jupiter. As it got close to Jupiter, the comet broke into 21 pieces. The pieces slammed into Jupiter for a week. Each impact created a crater larger than Earth.

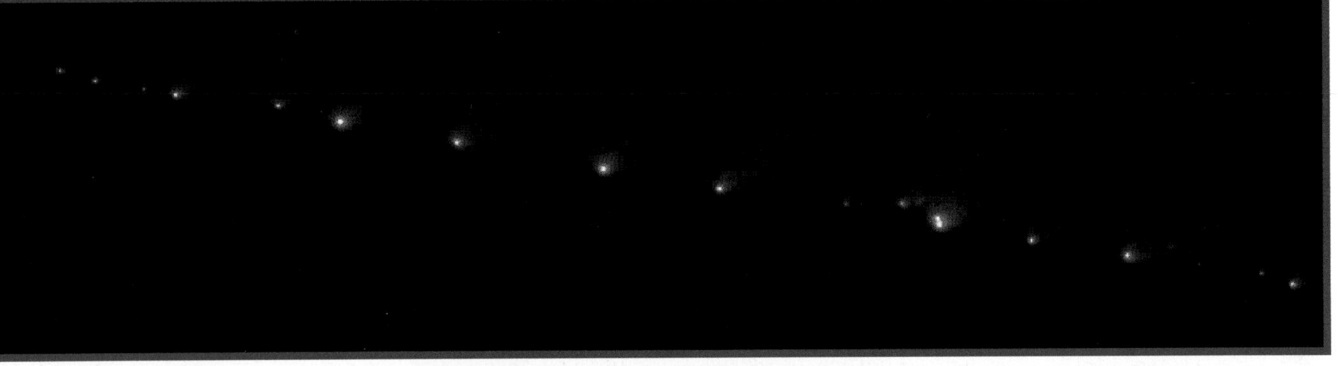

Comet Shoemaker-Levy 9 broke into 21 pieces as it got close to Jupiter.

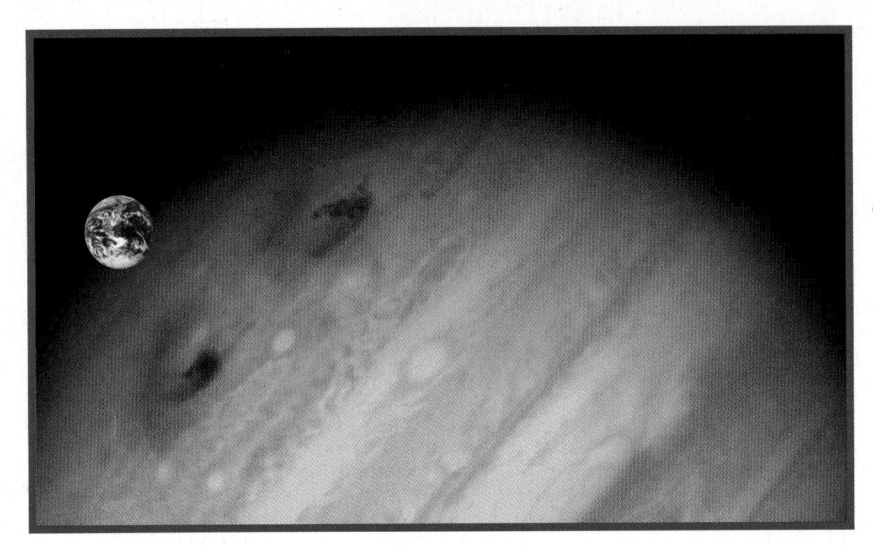

Two of the comet's craters on Jupiter. The picture of Earth gives an idea of how big the craters are.

Thinking about the Solar System

1. What is the Sun, and what is it made of?
2. What is the solar system?
3. Which planets are terrestrial planets? Which planets are gas giant planets?
4. What is the Kuiper Belt, and what is found there?
5. Which planet has the most moons orbiting it?
6. How are asteroids and comets alike and different?

Planets of the Solar System

Why Doesn't Earth Fly Off into Space?

Earth travels around the Sun in a predictable, almost-circular path once every year. That's a distance of about 942 million kilometers (km) each year. That's an incredible 2.6 million km each day! Earth travels at a speed over 100,000 km per hour. That's fast.

One important thing to know about objects in motion is that they travel only in straight lines. Objects don't change direction or follow curved paths unless a **force** pushes or pulls them in a new direction. If nothing pushed or pulled on Earth, it would fly off into space in a straight line.

But Earth doesn't fly off into space in a straight line. Earth travels in an almost-circular path around the Sun. In order to travel a circular path, Earth has to change direction all the time. Something has to push or pull Earth to change its direction. What is pushing or pulling Earth? The answer is gravity.

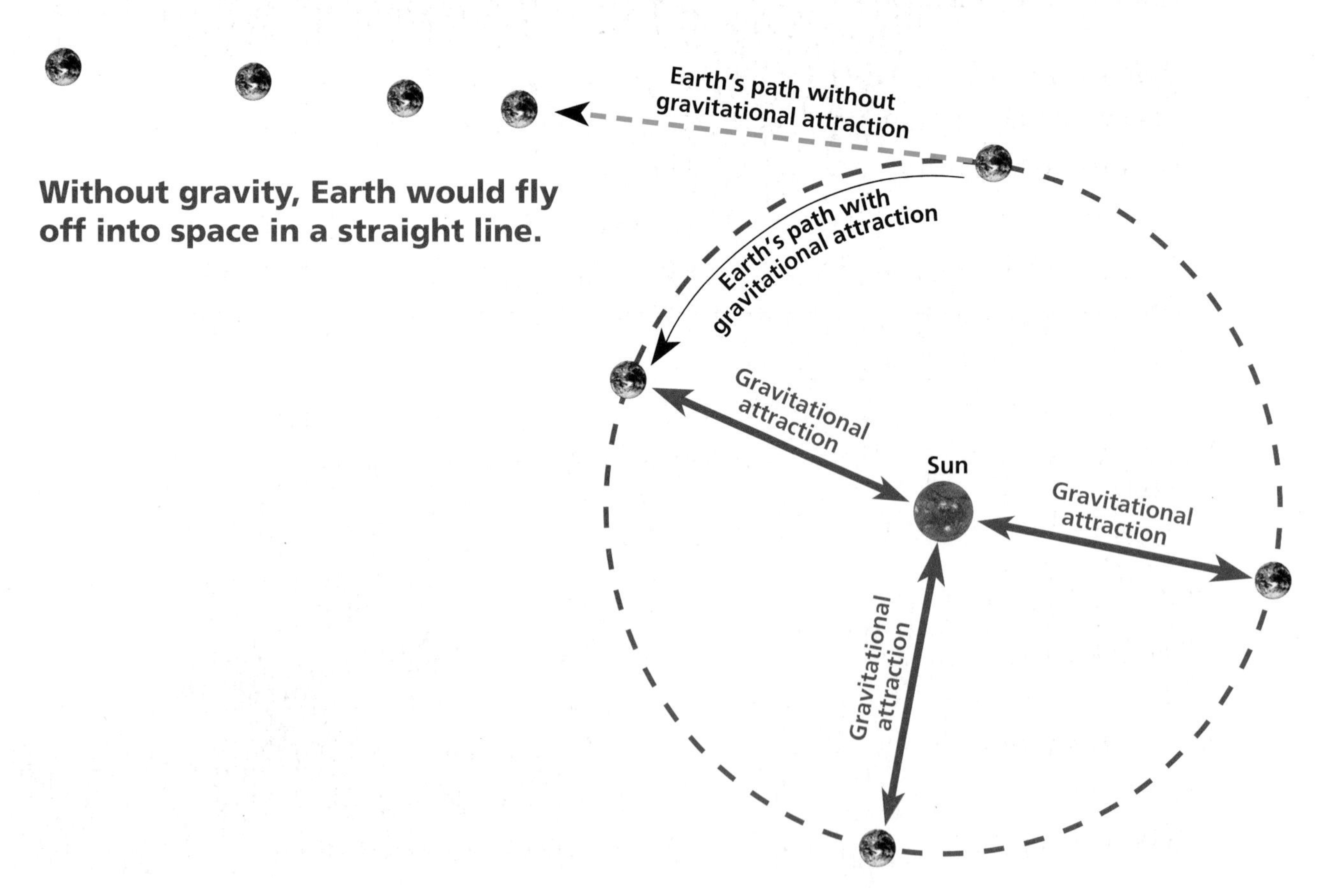

Without gravity, Earth would fly off into space in a straight line.

Gravity

Think about a ball resting motionless on a table. A gentle push on the ball will put it into motion. The ball will roll across the table. What will happen when the ball comes to the edge of the table? The ball will roll off the edge and fall to the ground. The ball's motion changes when it rolls off the edge of the table. It moves in a different direction and starts to move faster.

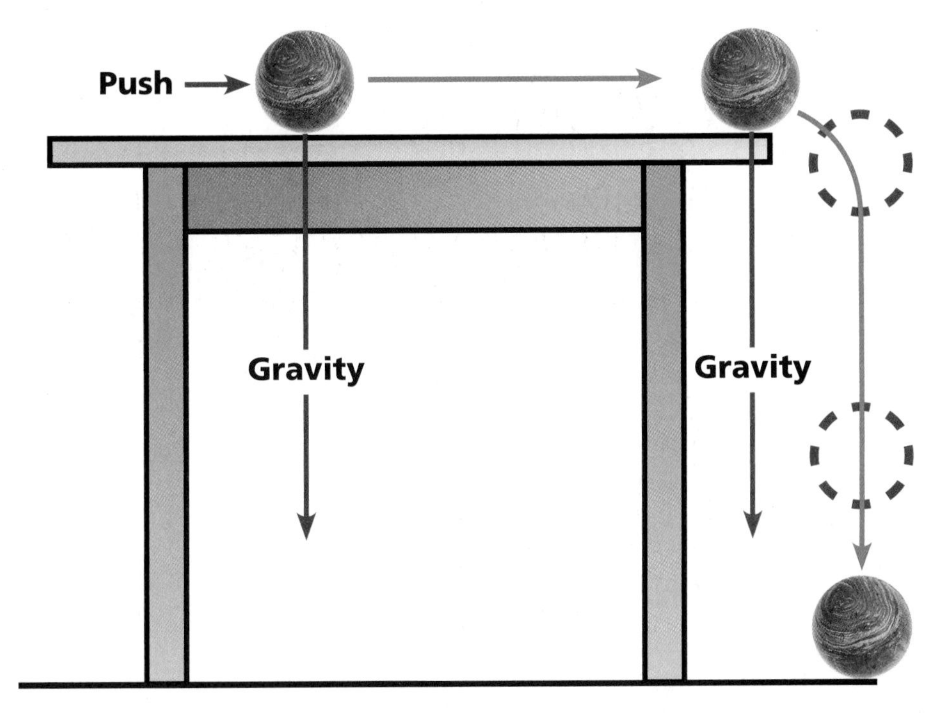

Gravity makes a ball fall to the ground.

What causes this change of motion? That's right, force. What force makes the ball move toward the ground? The force that makes the ball fall to the ground is gravity. Gravity is a pulling force between two objects, and it draws them toward each other. The bigger the objects, the stronger the gravitational force between them. Earth is a huge object, so it pulls strongly on all other objects. It is the force of gravity that pulls objects to Earth's center.

When you toss a basketball through a hoop, gravity pulls the ball to the ground.

If you return the ball to the flat tabletop, it will again rest there motionless. Why doesn't the ball fall to the ground? The ball doesn't move because the forces acting on it are balanced. There are two forces. First, the table is pushing upward on the ball. Second, gravity is pulling the ball downward toward Earth's center. When two equal forces act on an object in opposite directions, the forces are balanced. When the forces acting on an object are balanced, the object's motion does not change.

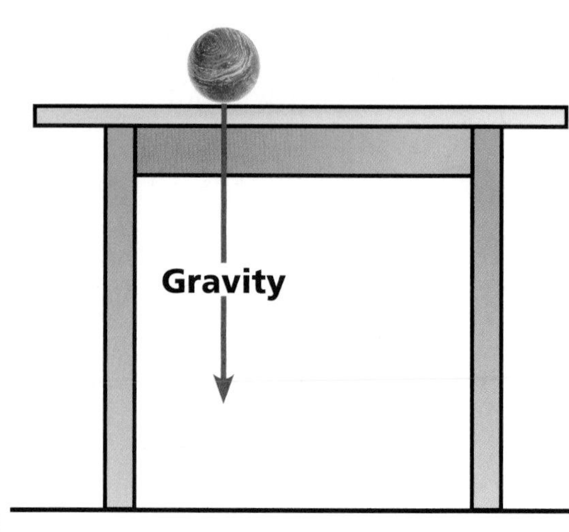

But what happens if you tip the table? The ball starts to roll down the table. For the ball to start moving, a force must act on the ball. Tipping the table unbalances the forces. The forces are no longer opposite and equal. Gravity pulls the ball downhill toward Earth's center. The round ball rolls across the table, over the edge, and down to the ground. When you're on a slide, can you feel the moment when the forces become unbalanced and gravity pulls you down?

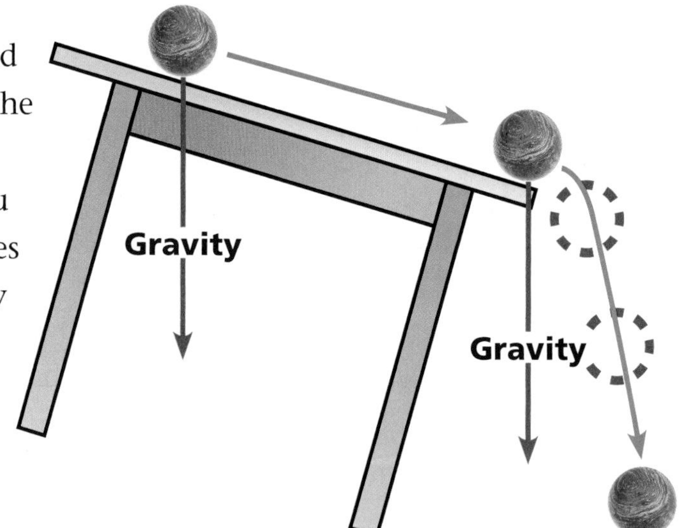

The force of gravity pulls the ball to the ground.

Gravity is the force of attraction between objects, or masses. The Sun is an object. Earth is an object. The force of attraction between the Sun and Earth pulls hard enough to change Earth's direction of travel.

Remember the string-and-ball demonstration? The hand pulled on the string. The string pulled on the ball. The ball traveled in a circular orbit. Gravity is like the string. The force of **gravitational attraction** between the Sun and Earth pulls on Earth, changing its direction of travel. The pull of gravity doesn't change Earth's speed, just its direction. That's why Earth travels in an almost-circular orbit around the Sun.

The Sun's gravity keeps all the planets in their orbits. Otherwise, each planet would fly off in a straight line right out of the solar system.

Earth travels around the Sun.

Thinking about Orbits

1. Why do planets stay in orbit around the Sun?
2. How is a ball on a string like a planet in its orbit?
3. What keeps the Moon in its orbit around Earth?

Stargazing

Stars are twinkling points of light in the night sky. When you get into bed at night, the sky is filled with stars. But in the morning, they are gone. Where did they go?

The stars didn't go anywhere. They are exactly where they were when you went to sleep. But you can't see the stars in the day sky. This is because the light from our star, the Sun, is so bright.

Where Are the Stars?

Stars are huge balls of hot gas. Most stars are located in groups of stars called galaxies. The Sun is in the galaxy called the Milky Way. There are several hundred billion other stars in the Milky Way with us. If we could see the entire Milky Way from above, it might look something like the picture below. The Sun is out on one of the arms where the arrow is pointing.

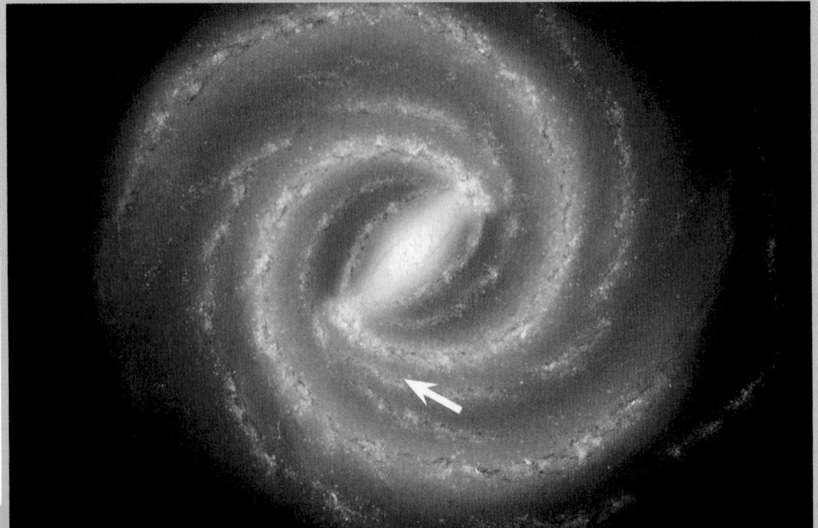

The Sun is one of the billions of stars in the Milky Way.

As you can see, we are surrounded by stars. Think about the 2,000 or so stars you can see and the billions of stars you can't see in the Milky Way. All these stars, including the Sun, are moving slowly around in a huge circle. Because all the stars move together, the positions of the stars never change. You can see the same stars in the same places in the sky year after year.

Did you ever see the **Big Dipper**? It is seven bright stars in the shape of a dipper. The Big Dipper is part of a **constellation** called Ursa Major, or the Great Bear.

Most of the stars you see in the night sky are part of a constellation. A constellation is a group of stars in a pattern. Thousands of years ago, stargazers imagined they could see animals and people in the star groups. They gave names to these constellations. Some of the names are Orion the hunter, Scorpius the scorpion, Aquila the eagle, Leo the lion, and Gemini the twins. Those same constellations are still seen in the sky today. They are unchanged.

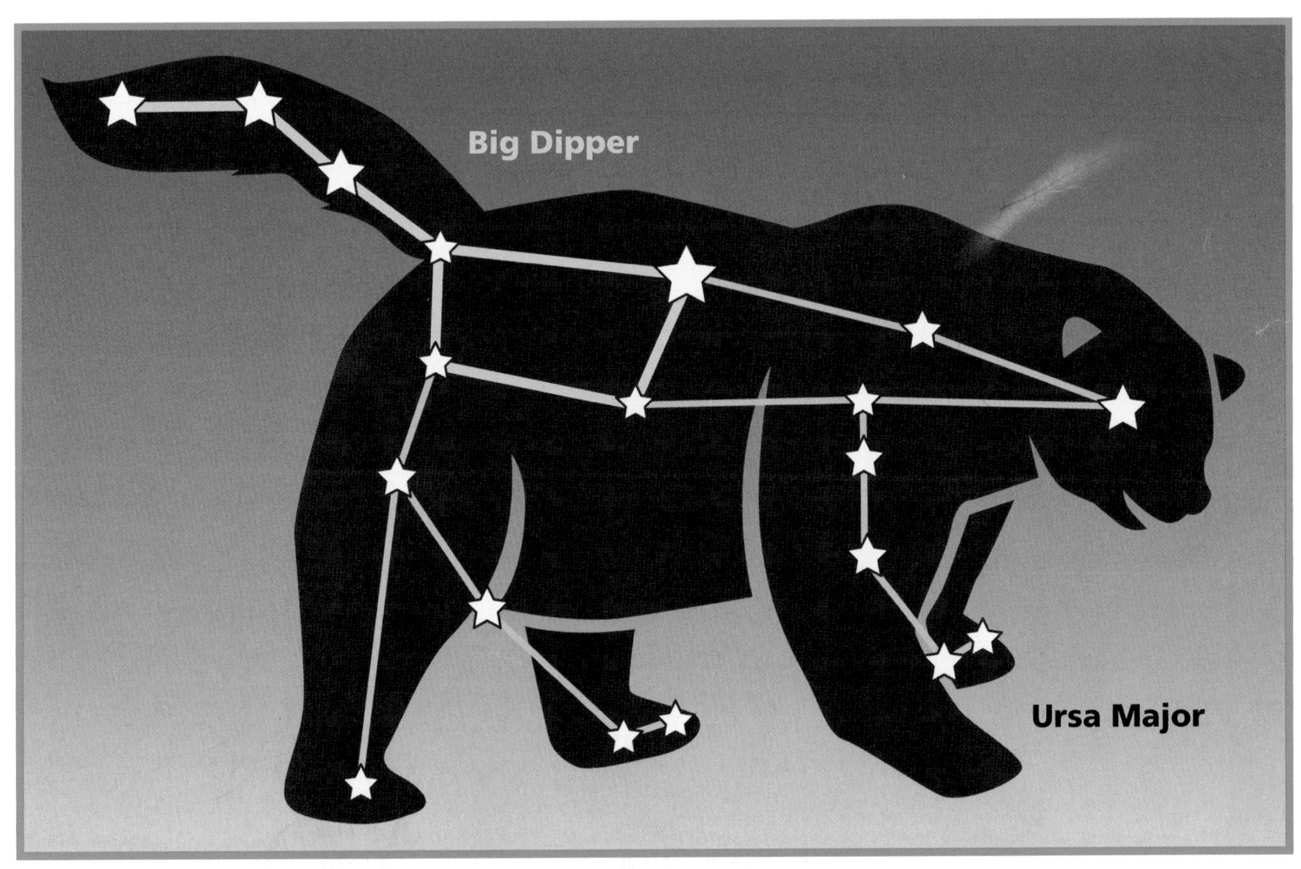

The constellation Ursa Major (the Great Bear)

Constellations in Motion

Even though the stars don't change position, they appear to move across the night sky. Stars move across the sky for the same reason that the Sun and the Moon move across the sky. The stars are not moving. Earth is moving. As Earth rotates on its **axis**, constellations rise in the east. They travel across the night sky and set in the west.

If you look at the stars every day for 1 year, you will see something interesting. The stars you see in winter are different from the stars you see in summer. If the stars don't move around, how is that possible? To answer this, we have to look at how Earth orbits the Sun.

Here is a simple drawing of the Milky Way. The Sun and Earth appear much larger than they really are. That's so we can study what happens as the seasons go by.

The side of Earth facing the Sun is always in daylight. The side facing away from the Sun is always in darkness. You can only see stars when you are on the dark half of Earth.

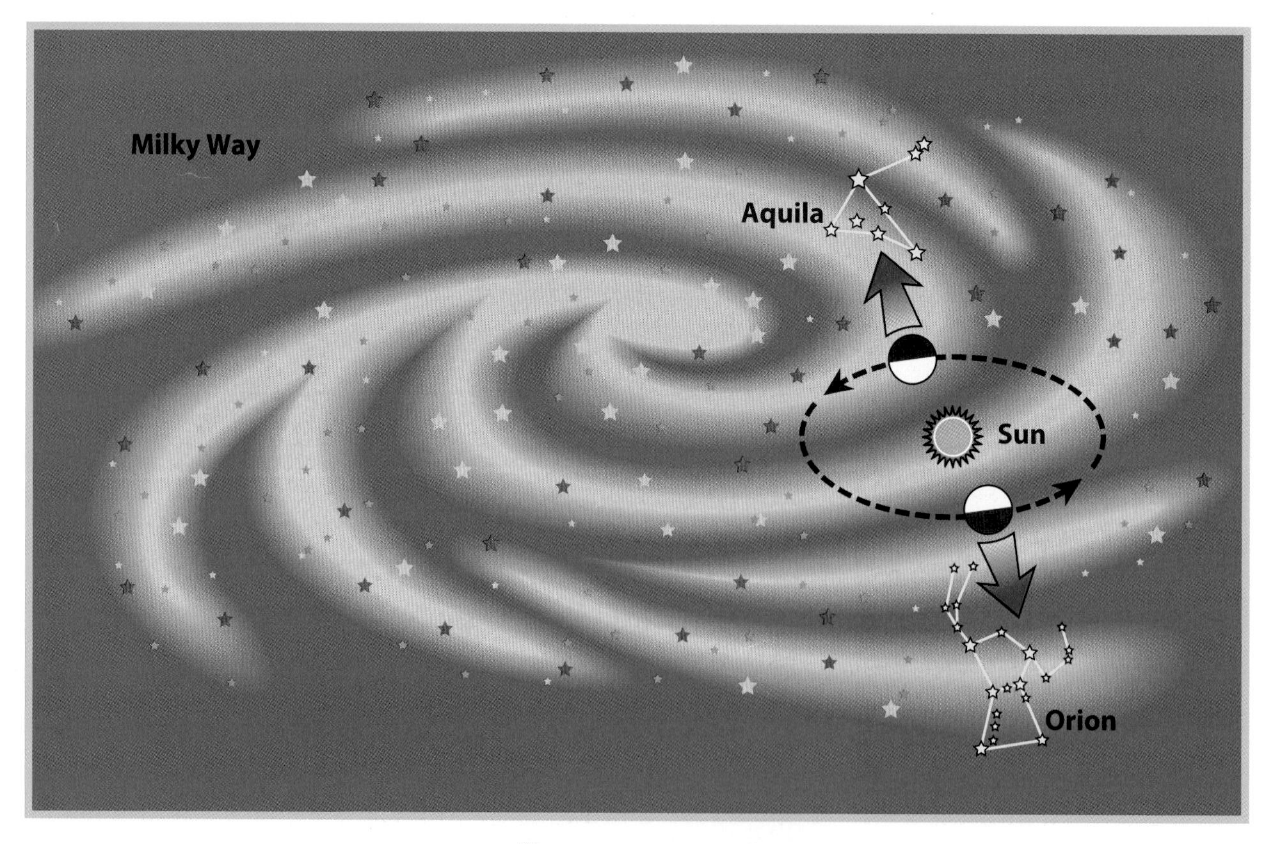

A simple drawing of the Sun and Earth, not drawn to scale.

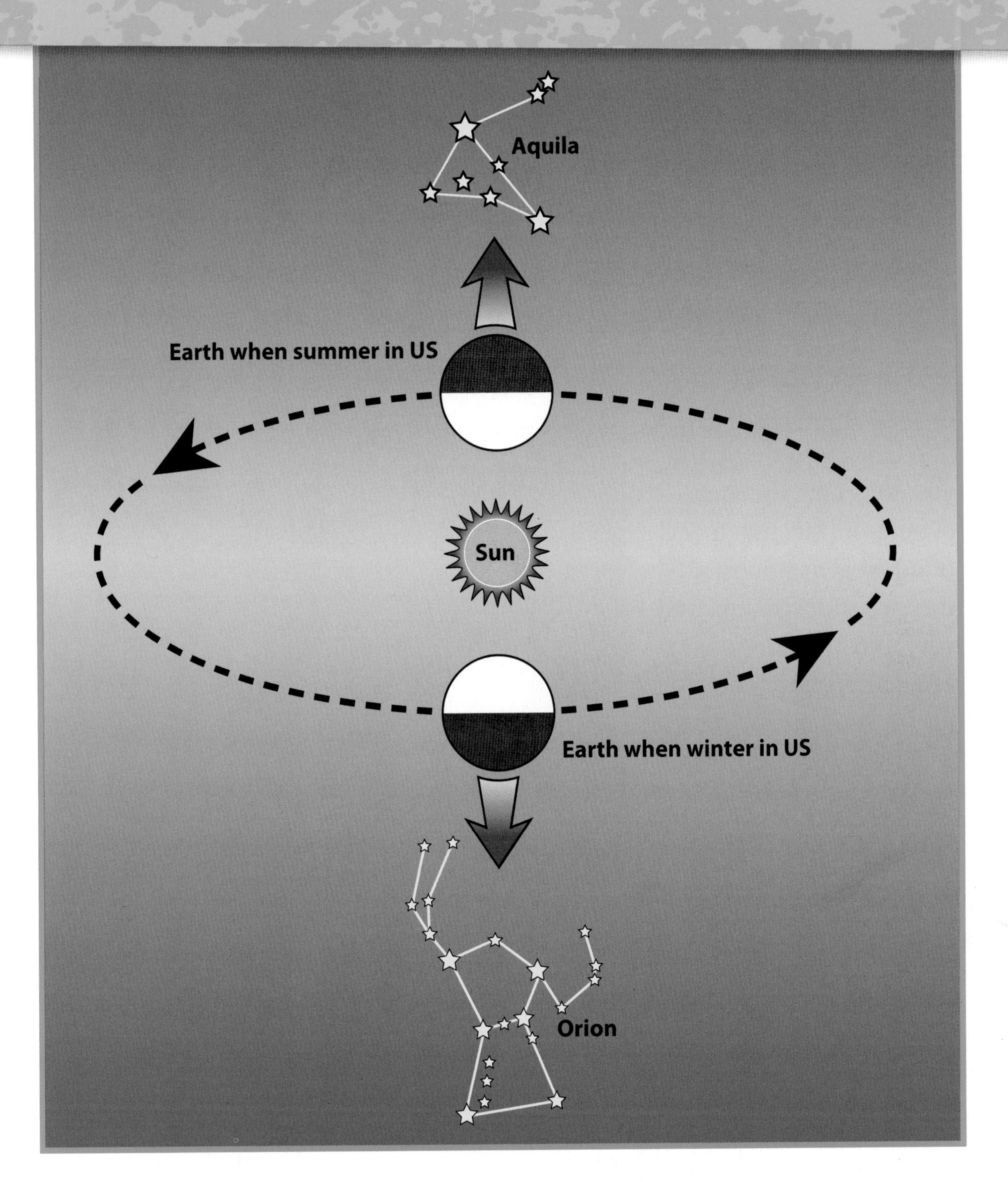

When it is summer in the United States, Earth is between the Sun and the center of the Milky Way. The constellation Aquila is in that direction. The dark side of Earth is toward the center of the galaxy in summer. On a summer night, you see Aquila high overhead.

Six months later, Earth is on the other side of the Sun. It is winter in the United States. Now the dark side of Earth faces away from the center of the galaxy. The constellation Orion is in that direction. On a winter night, you see Orion high overhead.

This is Orion. Can you see his belt and sword? The brightest stars in the Orion constellation appear in this picture. You can see Orion in the sky on a clear winter night.

When you see Orion, you are seeing the same pattern of stars that a hundred generations of stargazers looked at before you. And a hundred generations into the future, stargazers will still see Orion marching across the winter sky.

The constellation Orion is visible in the winter sky.

Thinking about the Stars

1. Why do stars appear to move across the night sky?
2. What is a constellation?
3. Why are the constellations in the summer sky different from those in the winter sky?
4. Imagine that you could see stars during the day. What constellation would you see at noon in winter? Why do you think so?

Star Scientists

Sometimes a childhood fascination with stars lasts a lifetime. Scientists who try to find out the secrets of stars are called astronomers. Meet five scientists who have taken star study in different directions. They truly are star scientists.

Stephen Hawking

When you drop a ball on Earth, the force of gravity pulls it down. Gravity keeps your feet on the ground, too. When a star reaches the end of its life, it collapses. Gravity pulls together all the matter in the star. In collapsed stars, gravity can even pull in light.

When a really big star collapses, it can become a **black hole**. In a black hole, gravity is so strong that nothing, not even light, can escape. Everything for millions of kilometers (km) around is pulled into the black hole, where it disappears.

Today, the best-known scientist who studies black holes is Stephen Hawking (1942–). Using mathematics, Hawking helped prove that black holes exist.

Stephen Hawking

Since 1994, the Hubble Space Telescope has been used to search for evidence of black holes. Hubble images show stars and gases swirling toward a central point. Hawking says this could be the effect of a black hole. A black hole's strong gravity would pull in everything around it, including stars. Future images from the Hubble Space Telescope and its successor, the James Webb Space Telescope, might help scientists improve their understanding of black holes.

Edna DeVore

Edna DeVore

Many people wonder if there is life anyplace else in the universe. But Edna DeVore (1947–) does more than wonder. DeVore is the Deputy Chief Executive Officer (CEO) of the SETI Institute. SETI is short for Search for Extraterrestrial Intelligence.

The scientists at SETI think that there might be other intelligent beings in the universe. If they are out there, they live on a planet orbiting a star. And intelligent life could develop technologies that send signals into space. Radio, TV, navigation systems, and telephones on Earth send messages in all directions into space. Is someone else out there doing the same thing?

The SETI Institute watches the sky for any signs of life in the universe. It uses big sets of antennae to listen for any sounds of life, like radio signals.

DeVore is a scientist and educator at SETI. She grew up on a ranch in Sattley, California. DeVore remembers watching the stars and the Milky Way in the clear night sky as a child. She didn't think about becoming a star scientist, but in college, DeVore became more and more interested in the stars. After getting her degree in **astronomy**, she became a teacher and a **planetarium** director. But the question she always asked herself was "Are we alone in the universe?"

DeVore is in charge of education and public outreach for the SETI Institute and NASA's Kepler Mission. And what's the latest report from the universe? The scientists at SETI haven't heard or seen anything yet. But they keep watching and listening.

SETI uses radio telescopes like this one.

Neil deGrasse Tyson

Neil deGrasse Tyson

You put water and fish in an aquarium. You put soil and plants in a terrarium. But what do you put in a planetarium? Planets! A planetarium is filled with information about planets, stars, galaxies, and everything else seen in the night sky.

A planetarium is a theater with a dome-shaped ceiling. In the middle of the room is a projector. The projector shines points of light all over the dome. The points of light are in the same positions as the stars in the sky. The projected stars make it seem as though you are outside watching the stars.

One of the fun things about a planetarium is that you can control the night sky. Do you want to see the stars as they were the day you were born? Or how the sky looked at different times in Earth's history? The projector operator can put you under the stars at any time and any place.

When he was a child, Neil deGrasse Tyson (1958–) never dreamed that he would one day be in charge of a planetarium. Tyson took a class at the Hayden Planetarium in New York City when he was in middle school. He was awarded a certificate at the end of the class. It meant a lot to him.

Tyson's love of the stars grew as he got older. After getting a PhD in astrophysics, Tyson spent time doing research and promoting education. He researched how stars form and explained space science to the public. He works hard to make science interesting for everyone. In 1996, Tyson became the youngest person ever to direct the Hayden Planetarium. It is the same place he visited as a child.

The Hayden Planetarium

Mae Jemison

Mae Jemison, astronaut

Mae Jemison (1956–) was born in Decatur, Alabama. She moved to Chicago, Illinois, as a child. There an uncle introduced her to astronomy. In high school, Jemison began reading books on astronomy and space travel. She was only 16 years old when she entered college. She earned degrees in chemical engineering and African and Afro-American studies from Stanford University. She went on to earn her medical degree from Cornell University.

After becoming a doctor, Jemison spent time in western Africa as a Peace Corps physician. But she continued to think about astronomy and space travel. She wanted to be part of the space program.

Jemison joined the astronaut program in 1987. On September 12, 1992, Jemison became the first African American woman in space. She was a science mission specialist on the space shuttle *Endeavour.* Jemison conducted experiments to find out more about the effects of being in space. She studied motion sickness, calcium loss in bones, and weightlessness.

The crew of *Endeavour*

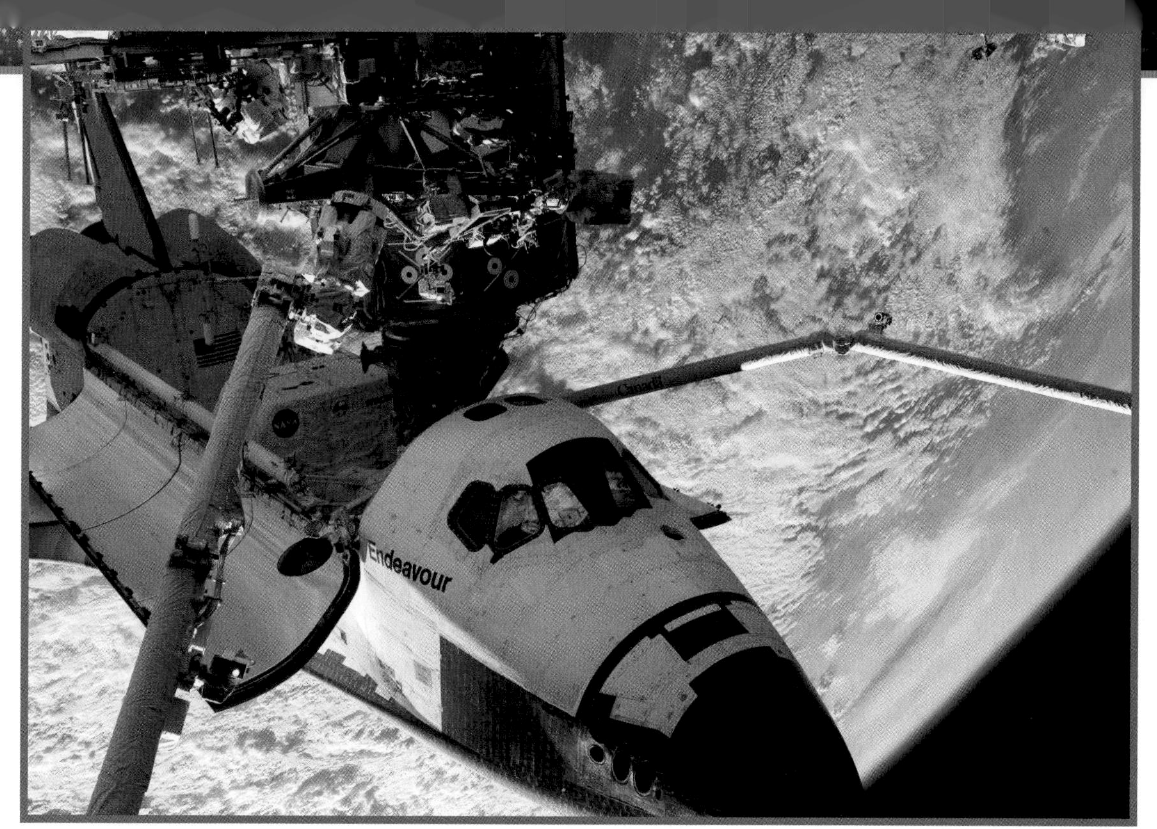

The space shuttle *Endeavour* docked at the International Space Station

Space shuttle mission STS-47 was the 50th space shuttle flight, but only the second flight for *Endeavour*. The space shuttle was in space for 8 days. During those 8 days, Jemison orbited Earth 127 times at an altitude of 307 km. The space shuttle traveled 5,234,950 km.

In 2011, after 30 years of flying and many firsts, the space shuttle program ended. Did the space shuttles actually fly in space? No, they orbited Earth in the upper atmosphere. What kept the shuttles in orbit? The answer is gravity. Shuttles traveled very fast. Earth's gravity pulled on the shuttles, constantly changing their direction of travel. Engineers from NASA figured out exactly how high above Earth's surface and how fast the shuttles needed to travel. Since they knew the force of gravity, the space shuttles were able to stay in orbit until the astronauts changed the speed. Then gravity pulled them back to Earth.

Ramon E. Lopez

Ramon E. Lopez

As strange as it may sound, there is **weather** in space. But it's not weather like we have on Earth. There are no clouds, **hurricanes**, or snowstorms in space. Space weather is the result of activities on the Sun. The Sun is always radiating energy into the solar system. The regular flow of light and gases is called **solar wind**. But what happens when the Sun goes through a period of violent solar flares? That's what Ramon E. Lopez (1959–) studies.

Solar flares are huge solar explosions. They send intense blasts of electrified gas into Earth's atmosphere. The blasts can produce electric effects in the atmosphere and on Earth's surface. The electricity can disable satellites orbiting Earth and interfere with radio transmissions and cell phones. Space weather can cause blackouts over large areas.

Lopez and his team understand how space weather can damage communication and navigation systems. And they understand how important these systems are to modern society. Can Lopez and his team learn how to predict space weather? Will it be possible to warn the world when a dangerous solar storm is coming? Lopez and the team he works with are developing a computer program to predict space weather about 30 minutes before it hits Earth. And that may be just long enough to protect communication and navigation systems from damage.

The Sun with a large solar flare

Our Galaxy

Stars are huge balls of hot gas. They produce bright light that travels out into space. When we go outdoors on a clear night, we see the stars as tiny points of light. There are billions of them sending light our way. But because most of them are so far away, the light is too dim for us to see. We can enjoy the 2,000 or so stars that we can see with our unaided eyes.

Astronomers study stars and other objects in the sky. One of the most important tools they use is the telescope. Telescopes **magnify** objects in the sky. When an astronomer looks at an object through a telescope, the object looks bigger and closer. With a telescope, you can see many more stars. Objects in the night sky can be studied in greater detail with a telescope.

The great Italian scientist Galileo Galilei used a telescope to observe the Moon and planets. He saw things no one had seen before. He observed mountains and craters on the Moon and discovered moons orbiting the planet Jupiter. Galileo's telescope brought the science of astronomy to a new level.

Galileo's painting of the phases of the Moon

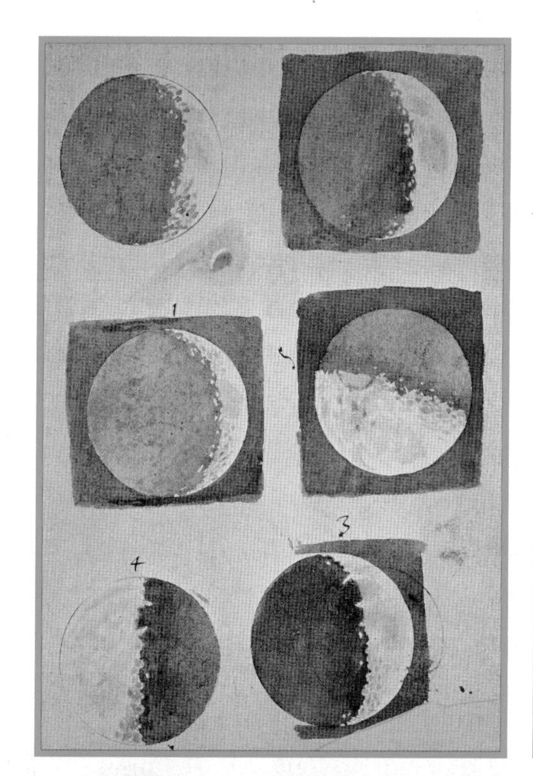

Galileo's record of the movement of Jupiter's moons

Moving Objects in the Sky

The Sun, the Moon, and the stars all move in the sky. But they all move in a different way. The Sun rises in the east in the morning and sets in the west at night. We see the Sun only during the day, never at night. Every day the Sun has the same pattern.

The Moon rises in the east and sets in the west, just like the Sun. But it doesn't always rise and set at the same time. Sometimes it rises in the morning, and sometimes it rises in the afternoon. There is a chance you might see the Moon in the daytime, but it might appear at night, depending on the lunar cycle.

The Moon and the Sun rise in the east and set in the west because Earth rotates on its axis. To people on Earth, the Sun and the Moon appear to move across the sky. The Moon rises and sets at different times because the Moon is orbiting Earth. The Moon is changing its position all the time.

Stars are different. We see them only at night. They are up in the sky all the time. But we can't see them during the day because the Sun is too bright. As soon as the Sun sets, the sky gets dark. Then we can see the stars. Stars rise and set, too. If you watch one star, you can see it rise above the eastern horizon. It then appears to move across the night sky, and set in the west. Why? Because Earth is rotating.

Earth's Orbit

One more thing is different about stars. Earth is completely surrounded by stars in all directions. But you can't see all of them at once. This is because half of them are on the day side of Earth. Also, the stars you can see in winter are different from the ones you can see in summer.

Here's why. Earth orbits the Sun. One complete orbit takes a year. At all times, half of Earth is in daylight and half is in darkness. It is always the side of Earth toward the Sun that is in daylight. The day side of Earth is always "looking" toward the Sun.

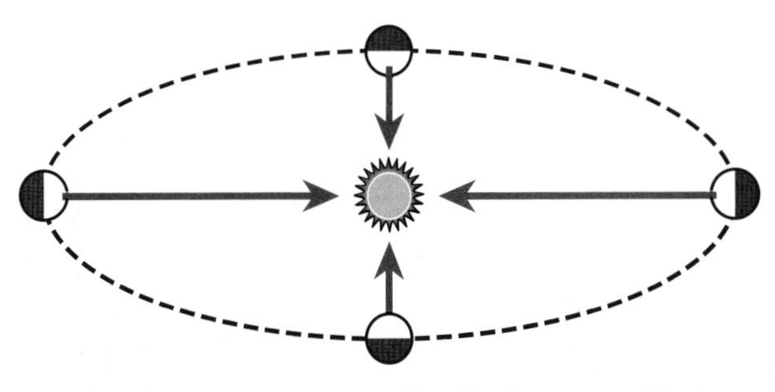

Half of Earth is in daylight at all times.

The side of Earth away from the Sun is always in darkness. The dark side of Earth is always "looking" away from the Sun. We can only see stars at night when it is dark. So stargazers always look in the direction away from the Sun.

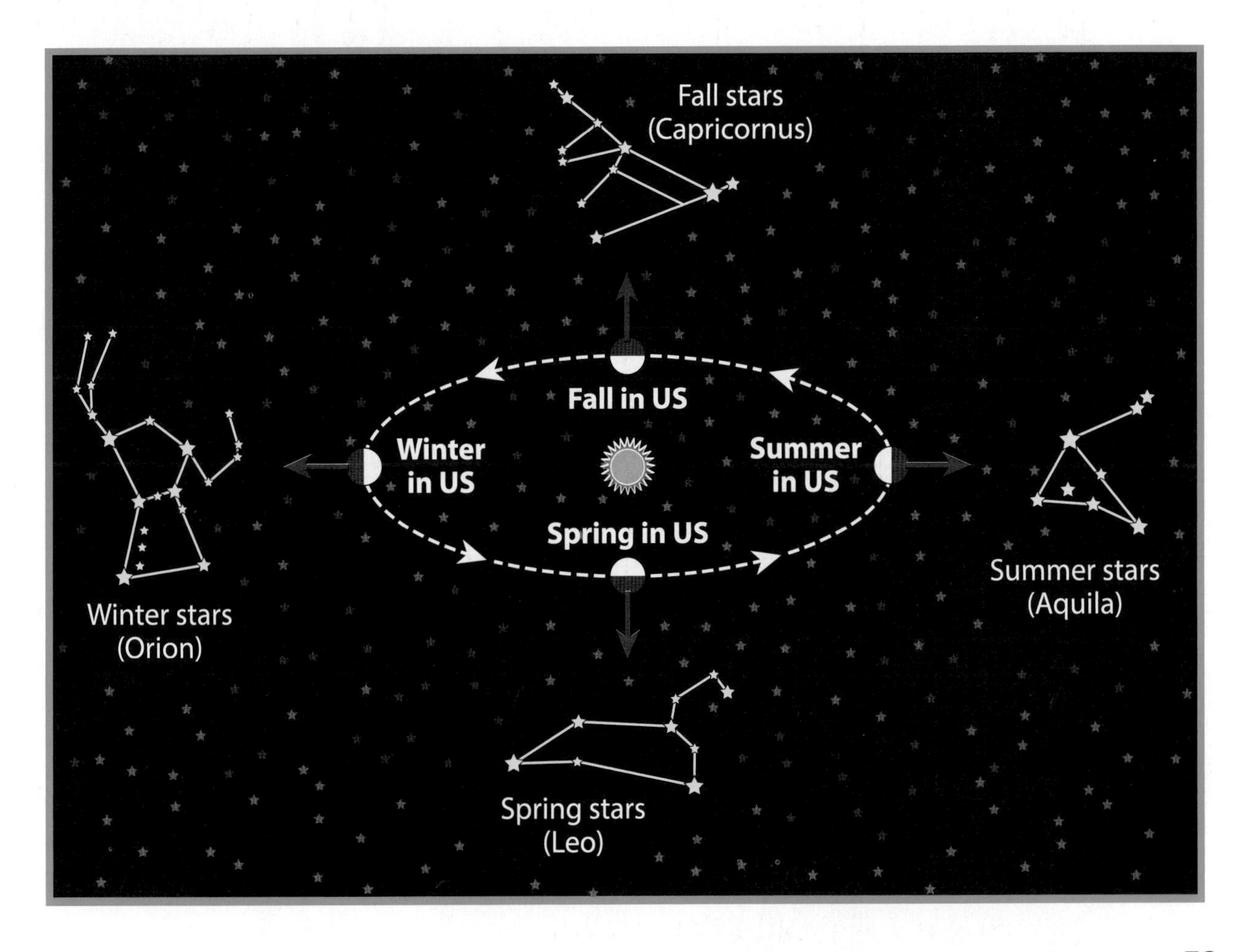

As Earth orbits the Sun, the dark side of Earth is aimed at different parts of the star-filled sky. Because stars don't move, the stars and constellations you see change from season to season.

In the United States, look for Aquila the eagle in summer. Look for Capricornus the goat in the fall, Orion the hunter in winter, and Leo the lion in spring.

Leo the lion

Clouds float in air.

What Is Air?

You can't see it or taste it. You can't even smell it. But you might feel it as a gentle breeze brushing across your skin. Air is difficult to understand because it is not easy to observe with your senses. Is air one thing or a mixture of things? And where is air? Is it everywhere or just in some places?

As we go about our everyday lives, we usually travel with our feet on solid Earth and our heads in the atmosphere. The atmosphere is all around us, pressing firmly on every part of our bodies—top, front, back, and sides. Even if we attempt to get out of the atmosphere by going inside a car or hiding in a basement, the atmosphere is there, filling every space we enter.

An atmosphere is the layer of gases surrounding a planet or star. All planets and stars have an atmosphere around them. The Sun's atmosphere is mostly hydrogen. Mars has a thin atmosphere of carbon dioxide (CO_2) with a bit of nitrogen and a trace of **water vapor**. Mercury has almost no atmosphere at all. Each planet is surrounded by its own mixture of gases.

A view from space, looking down through Earth's atmosphere

Earth's atmosphere is made up of a mixture of gases we call air. Air is mostly nitrogen (78 percent) and oxygen (21 percent), with some argon (0.93 percent), carbon dioxide (0.039 percent), **ozone**, water vapor, and other gases (less than 0.04 percent together).

Nitrogen is the most abundant gas in our atmosphere. It is a stable gas, which means it doesn't react easily with other substances. When we breathe air, the nitrogen goes into our lungs and then back out unchanged. We don't need nitrogen gas to survive.

Oxygen is the second most abundant gas. It makes up about 21 percent of air's volume, and it accounts for 23 percent of air's mass. Oxygen is a colorless, odorless, and tasteless gas. Oxygen combines with hydrogen to form water. Without oxygen, life as we know it would not exist on Earth.

Oxygen and nitrogen are called permanent gases. The amount of oxygen and nitrogen in the atmosphere stays constant. The other gases in the table are also permanent gases, but are found in much smaller quantities in the atmosphere.

Permanent Gases of the Atmosphere

Gas	Percentage by volume
Nitrogen	78.08%
Oxygen	20.95%
Argon	0.93%
Neon	0.002%
Helium	0.0005%
Krypton	0.0001%
Hydrogen	0.00005%
Xenon	0.000009%

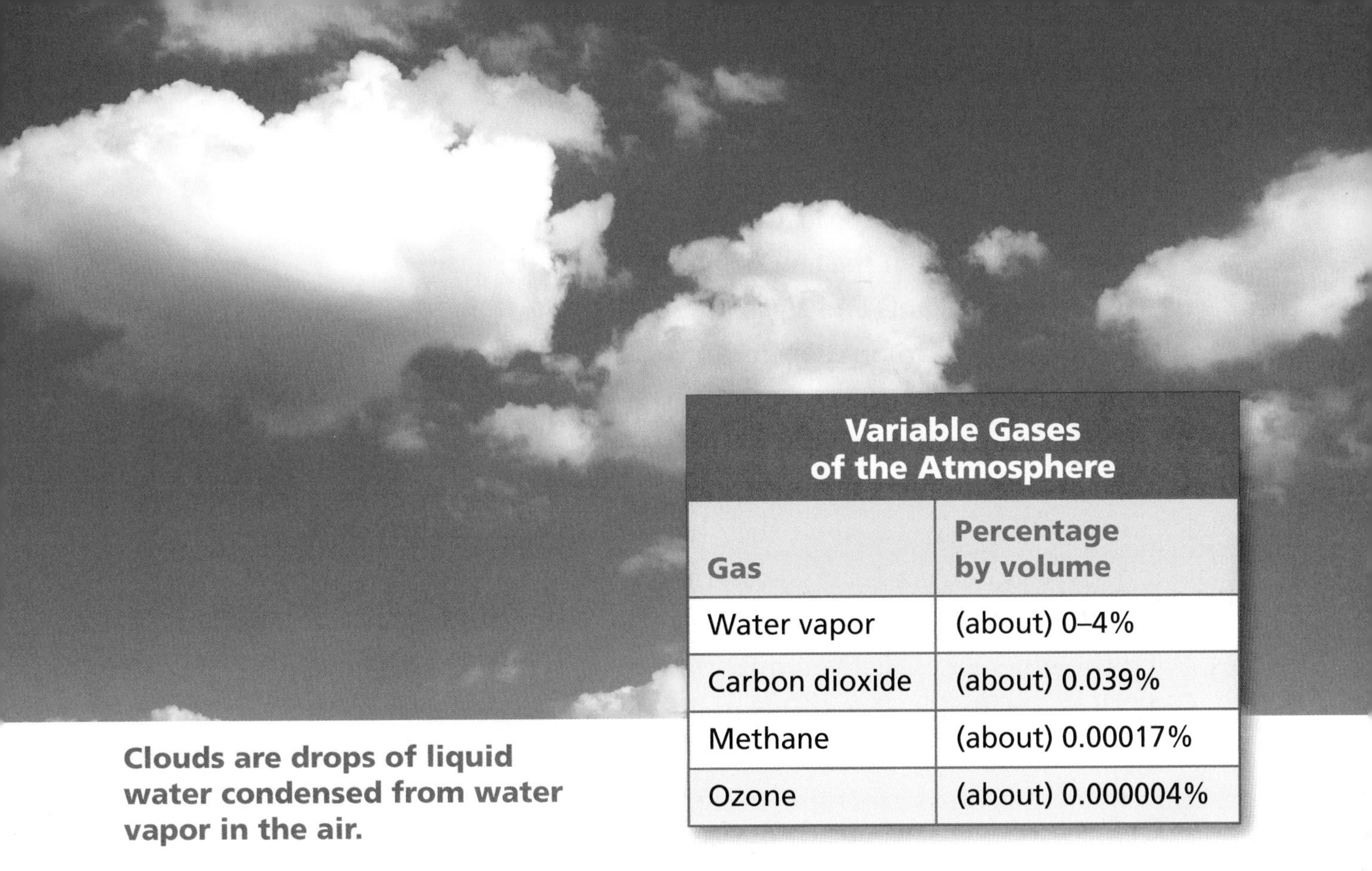

Variable Gases of the Atmosphere	
Gas	Percentage by volume
Water vapor	(about) 0–4%
Carbon dioxide	(about) 0.039%
Methane	(about) 0.00017%
Ozone	(about) 0.000004%

Clouds are drops of liquid water condensed from water vapor in the air.

Air also contains variable gases. The amount of each variable gas changes in response to activities in the environment.

Water vapor is the most abundant variable gas. It makes up about 0.25 percent of the atmosphere's mass. The amount of water vapor in the atmosphere changes constantly. Water moves between Earth's surface and the atmosphere through **evaporation**, **condensation**, and **precipitation**. You can get an idea of the changes in atmospheric water vapor by observing clouds and noting the stickiness you feel on your skin on humid days.

Carbon dioxide is another important variable gas. It makes up only about 0.039 percent of the atmosphere. You can't see or feel changes in the amount of carbon dioxide in the atmosphere.

Precipitation, such as rain and snow, comes from water vapor in the air.

Carbon dioxide plays an important role in the lives of plants and algae. Carbon dioxide is removed from the air during **photosynthesis**. Plants and algae use light from the Sun, carbon dioxide, and water to produce sugar (food). During this process, they release oxygen to the atmosphere. When living organisms use the energy of food to stay alive, they remove oxygen from the air and return carbon dioxide to the air.

Plants make sugar out of sunlight, carbon dioxide, and water.

Here are other gases that you might have heard about. Methane is a variable gas that is increasing in concentration in the atmosphere. Scientists are trying to figure out why this is happening. They suspect several things. Cattle produce methane in their digestive processes. Methane also comes from coal mines, oil wells, and gas pipelines. It is a by-product of rice cultivation and melting permafrost in arctic regions. Methane **absorbs** energy and transfers heat to the atmosphere.

Ozone is a variable gas. It is a form of oxygen that forms a thin layer high in the atmosphere. The ozone layer protects life on Earth by absorbing dangerous ultraviolet (UV) light from the Sun. But ozone in high concentrations can cause lung damage. In the lower atmosphere, ozone is an air pollutant.

Some methane gas comes from the digestive processes of cows.

Thinking about Air

What is air?

Earth's Atmosphere

Earth's atmosphere is made up of a mixture of permanent and variable gases. These gases are all mixed together. Any sample of air is a mixture of all of them. The gases mix because the air particles are always moving near Earth's surface. Above about 90 kilometers (km), the gases mix less, and there are more light gases, such as hydrogen and helium.

There are extreme temperatures in the universe. The temperature can be as cold as –270 degrees Celsius (°C). Near hot stars, such as the Sun, it can be very hot, up to thousands of degrees. But there are a few places in the universe that have a temperature between those extremes of hot and cold. Earth is one of those places where the temperature is just right.

On a typical day, the temperature range on Earth is only about 100°C. It might be 45°C in the hottest place on Earth and –55°C at one of the poles. The measured extremes are 57°C in Death Valley, California, recorded on July 10, 1913, and –89°C in Vostok, Antarctica, on July 21, 1983. That's a range of temperature on Earth of 147°C.

Space-shuttle astronauts took this photo while orbiting Earth. You can see a side view of Earth's atmosphere. The black bumps pushing into the troposphere are tall cumulus clouds.

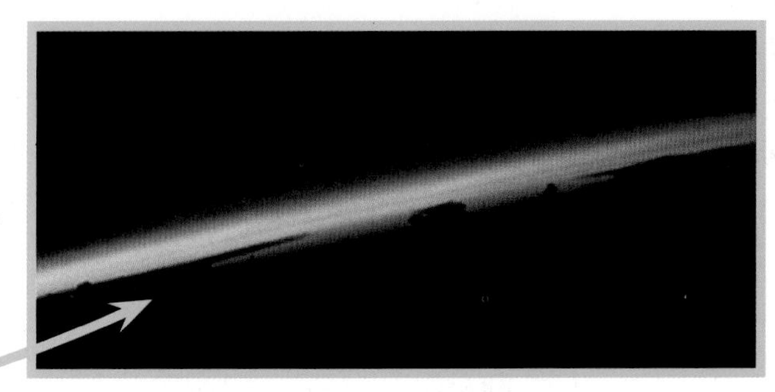

The crew of Apollo 17 took this photo of Earth in December 1972, while on their way to the Moon. The small green box at the top of Earth's image shows about how much area is in the atmosphere photo.

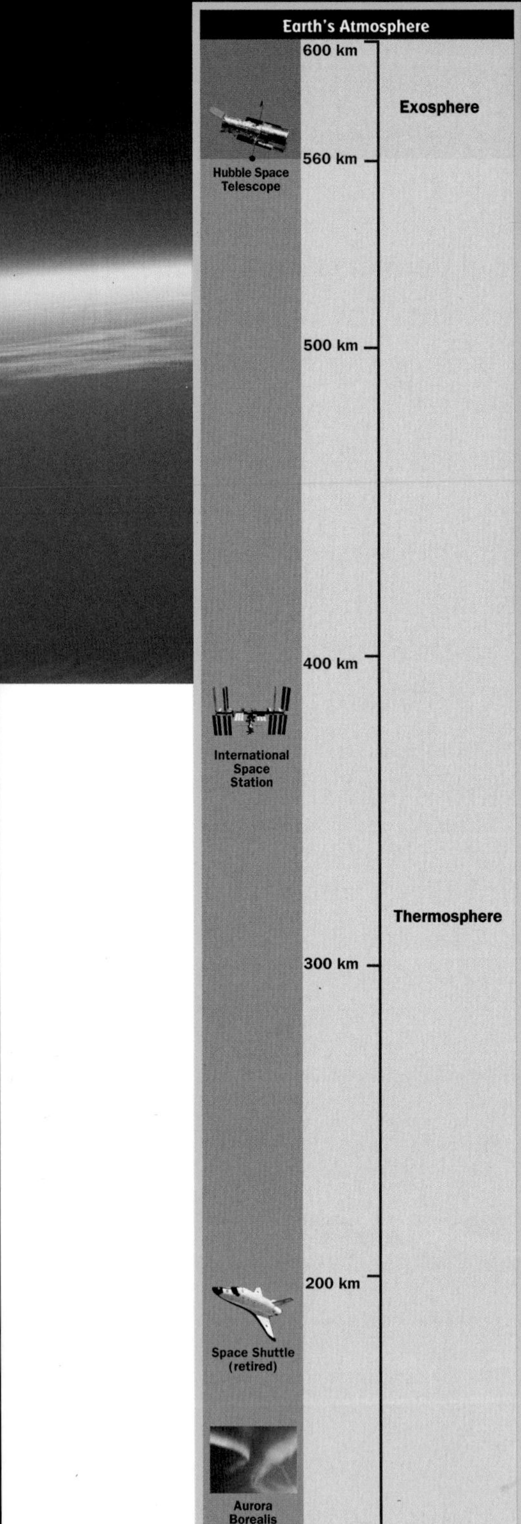

Earth's atmosphere looks like a thin, blue veil.

It's not only because we are at the right distance from the Sun that Earth's temperatures are moderate. Earth's atmosphere keeps the temperature within a narrow range so that it is just right for life on Earth.

From space, Earth's atmosphere looks like a thin, blue veil. Some people like to think of the atmosphere as an ocean of air covering Earth. The depth of this "ocean" is about 600 km. The atmosphere is most dense right at the bottom, where it rests on Earth's surface. It gets thinner and thinner (less dense) as you move away from Earth's surface. There is no real boundary between the atmosphere and space. The air just gets thinner and thinner until it disappears.

Imagine a column of air that starts on Earth's surface and extends up 600 km to the top of the atmosphere. Scientists have discovered several layers in this column of air. Each layer has a different temperature. Here's how it stacks up.

The four seasons occur in the troposphere.

The layer we live in is the **troposphere**. It starts at Earth's surface and extends upward for 9–20 km. Its thickness depends on the season and where you are on Earth. Over the warm equator, the troposphere is a little thicker than it is over the polar regions, where the air is colder. It also thickens during the summer and thins during the winter. The average thickness of the troposphere is 15 km.

This ground-floor layer has most of the organisms, dust, water vapor, and clouds found in the entire atmosphere. It has most of the air as well. And, most important, weather occurs in the troposphere. The troposphere is where differences in air temperature, **humidity**, **air pressure**, and **wind** occur.

These properties of temperature, humidity, air pressure, and wind are important **weather variables**. **Meteorologists** launch weather balloons twice each day to monitor weather variables. The balloons float up through the troposphere to about 18 km.

The troposphere is the thinnest layer. It has only about 2 percent of the depth of the atmosphere. It is the most dense layer, however, containing four-fifths (80 percent) of the total mass of the atmosphere.

Earth's surface (both land and water) absorbs heat from the Sun and warms the air above it. Because air in the troposphere is heated mostly by Earth's surface, the air is warmest close to the ground. The air temperature drops as you go higher. At its highest point, the temperature of the troposphere is about –60°C. The average temperature of the troposphere is about 25°C.

Mount Everest, located in Nepal and Tibet, is the highest landform on Earth, rising 8.848 km into the troposphere. The air temperature at the top of the mountain is far below freezing most of the time. There is also less air to breathe at the top of Mount Everest. Climbers usually bring oxygen to help them survive the thin air.

Mount Everest

Jets usually fly in the region between the lower stratosphere and the upper troposphere.

The **stratosphere** is the layer above the troposphere. It is 15–50 km above Earth's surface and has almost no moisture or dust. It does, however, have a layer of ozone that absorbs ultraviolet (UV) **radiation** from the Sun. The temperature stays below freezing until you reach the top part of the stratosphere, where ozone absorbs energy and warms the air to about 0°C.

The jet stream, a fast-flowing stream of wind, travels generally west to east in the region between the lower stratosphere and the upper troposphere. Many military and commercial jets take advantage of the jet stream when flying from west to east. The jet stream winds move cold air over North America. This brings cold temperatures and storms.

The **mesosphere** is above the stratosphere, 50–85 km above Earth's surface. The temperature is colder than in the stratosphere. Its coldest temperature is around –90°C in the upper mesosphere. This is the layer where meteors burn up while entering Earth's atmosphere. We call these burning meteors shooting stars.

Beyond the mesosphere, 85–560 km (or higher) above Earth, is the **thermosphere**. The thermosphere is the most difficult layer of the atmosphere to measure. The air is extremely thin. The thermosphere is where the atmosphere is first heated by the Sun. A small amount of energy coming from the Sun can result in a large temperature change. When the Sun is very active with sunspots or flares, the temperature of the thermosphere can be 1,500°C or higher!

Temperature defines these four layers. The boundaries between the layers are not fixed lines and they can change with the seasons.

Meteors burn up in the mesosphere.

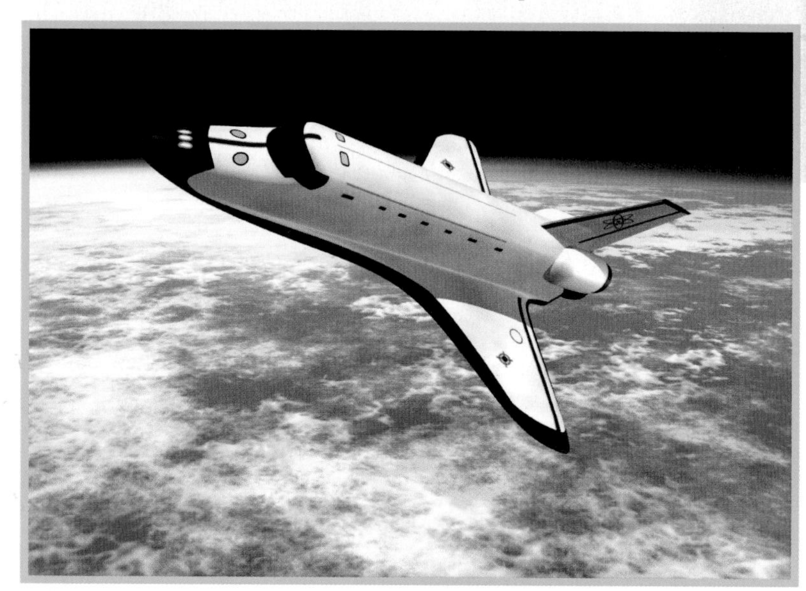

The space shuttle orbited Earth in the thermosphere.

The International Space Station is in the thermosphere above Earth.

Earth's atmosphere fades into space.

Beyond the thermosphere, Earth's atmosphere fades into space. The **exosphere** is where gas particles escape into space. In this region, the temperature drops to the extreme –270°C of outer space.

That 600 km column of air pushes down on the surface of Earth with a lot of force. We call the force air pressure. We are not aware of it because we are adapted to live under all that pressure. But there is a mass of about 1 kilogram (kg) pushing down on every square centimeter (cm) of surface on Earth. Your head has a surface area of about 150 square cm. This means you have about 150 kg of air pushing down on your head. That's like having a kitchen stove or a motorcycle pushing down on your head all the time!

Here's another way to look at it. If all the air were replaced with solid gold, the entire planet would be covered by a layer of gold about half a meter deep. The mass of the entire atmosphere is about the same as half a meter of gold. But the atmosphere is much more valuable.

Thinking about the Atmosphere

1. How is Earth's atmosphere like the ocean? How is it unlike the ocean?
2. What is the average temperature of the troposphere? Why is that important?

Weather Instruments

Meteorologists are scientists who study weather. Weather is the condition of the air in an area. The conditions can change, so they are called weather variables. The most important weather variables to meteorologists are temperature, air pressure, humidity, and wind. Meteorologists use weather instruments to measure each variable.

A weather tower with a weather station on top

Temperature

Temperature is a measure of how hot the air is. Temperature is measured with a **thermometer**. There are many kinds of thermometers. The most common kind is a liquid thermometer. A liquid thermometer is a thin glass tube connected to a small bulb of liquid. As the liquid warms and cools, it expands and contracts. The height of the column of liquid in the tube changes in response to the temperature. By labeling the liquid tube to show temperatures, the meteorologist can read the temperature directly from the thermometer.

Metals also expand and contract in response to temperature change. Some thermometers use strips made of two different metals to detect temperature changes. These are called bimetallic thermometers. The two metals have different rates of expansion. One side of the strip expands more than the other as it heats up, and the strip bends. A pointer on the end of the bending strip points to the temperature.

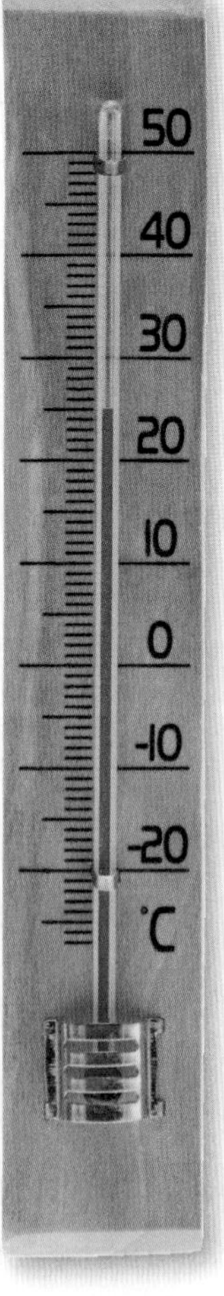

A liquid thermometer

A bimetallic thermometer

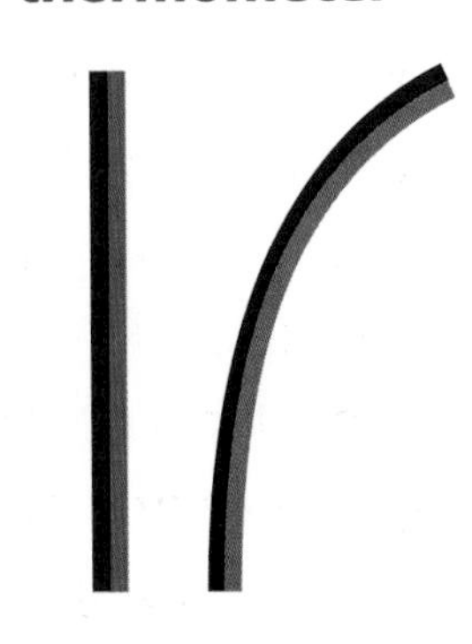

cold hot

Air Pressure

Air pressure is the force of air pushing on things around it. Air pressure changes with the density of the air. When air heats up, it becomes less dense; when it cools, it becomes more dense. The instrument that measures air pressure is called a **barometer**. Air pushes on a closed container, one side of which is attached to a dial in the barometer. The harder the air pushes, the higher the dial goes. The dial measures in units called millibars. Changes in air pressure mean that weather conditions will change. Falling air pressure means **rain** is coming. Rising air pressure means fair and dry weather is coming.

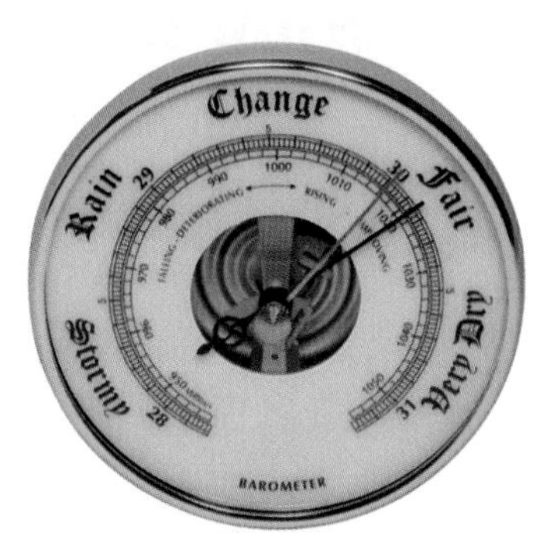

A barometer

Humidity

Water vapor is water (H_2O) in its gas state. As vapor, water can enter the air. The water vapor will eventually condense and form drops of water, which can fall as rain. Meteorologists measure humidity, the amount of water in the air, with instruments called **hygrometers**. Humidity is measured as a percentage.

A hygrometer

Wind Speed

Moving air is called wind. Meteorologists are interested in how fast the wind is moving. To measure wind speed, meteorologists use **anemometers** and **wind meters**. An anemometer uses a rotating shaft with wind-catching cups attached at the top. The harder the wind blows, the faster the shaft rotates, and the faster the cups move through the air. The moving cups measure the wind speed.

A wind meter is an instrument with a small ball in a tube. When wind blows across the top of the tube, the flow of air up the tube lifts the ball. The harder the wind blows, the higher the ball rises. Both instruments are adjusted to report wind in miles per hour (mph) or kilometers (km) per hour.

An anemometer

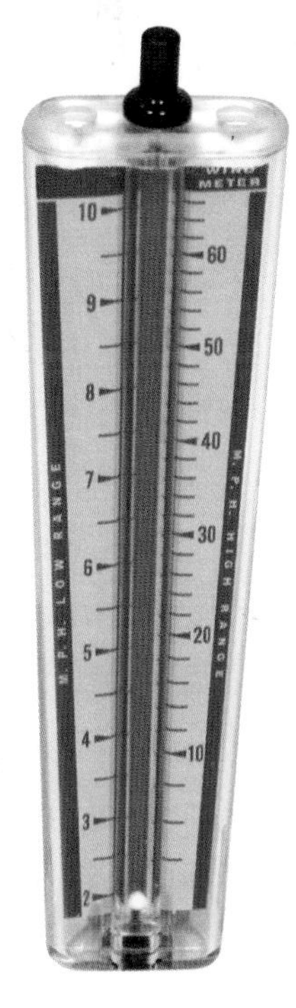

A wind meter

Wind Direction

Meteorologists are also interested in the direction the wind is blowing. To determine wind direction, meteorologists use a **wind vane**. A wind vane is a shaft with an arrow point on one end and a broad paddle shape at the other end. When wind hits the paddle, it rotates the shaft so that the arrow points into the wind. Using a compass, the meteorologist determines the direction the shaft is pointing. Wind direction is the direction from which the wind is blowing. It is reported in compass directions, such as north or south.

A wind vane

A compass

Modern Weather Instruments

Meteorologists now use a combination of traditional weather instruments and computer-based digital weather instruments. Meteorologists get information from advanced electronic instruments that are placed in good locations for monitoring weather. Those instruments use radio transmitters (like those in cell phones) to send information to weather centers where meteorologists work.

This weather device for home use has electronic instruments inside for detecting and reporting temperature and humidity. Some models measure air pressure and are connected to anemometers to measure wind speed.

Uneven Heating

Stars create a lot of energy. Energy radiates from them in all directions. Most of this energy streams out into space and never hits anything. A small amount, however, hits objects in the universe. When you look into the sky on a dark, clear night, you see thousands of stars. You see them because a tiny amount of energy from the stars goes into your eyes.

During the day, you are aware of the energy coming from a much closer star, the Sun. When sunlight comes to Earth, you can see the light and feel the heat when your skin absorbs the light. Heat and light from the Sun are called **solar energy**.

When light from the Sun hits matter, such as Earth's surface, two important things can happen. The light can be reflected or absorbed by the matter. If the light is reflected, it simply bounces off the matter and continues on its way in a new direction. But if the light is absorbed, the matter gains energy. Usually the gained energy is heat. When matter absorbs energy, its temperature goes up.

Heating It Up

The amount of solar energy coming from the Sun is about the same all the time. But the temperature of Earth's surface is not even. Some locations get warmer than other locations. Why is that?

There are several variables that affect how hot a material will get when solar energy shines on it. The table below lists several variables and how each affects the temperature change of a material.

Variable	Effect
Length of exposure	Longer exposure leads to higher temperature.
Intensity of solar energy	Greater intensity leads to higher temperature.
Angle of exposure	More direct angle leads to higher temperature.
Color of material	Darker color leads to higher temperature.
Properties of material	Water shows the least temperature change.

Length of exposure is how long the Sun shines on an object.

Intensity of solar energy is how bright and concentrated the energy is. For example, if the light travels through clouds, it will be less intense. Clouds reflect and absorb some of the energy before it gets to Earth's surface. The brighter the sunlight falling on an object, the warmer the object will get.

Angle of exposure changes throughout the day. Morning sunshine comes at a low angle and is less intense and weak. Midday sunshine radiates down from a high angle and is intense and strong.

Different colors absorb solar energy differently. Black absorbs the most solar energy. White absorbs the least solar energy.

The chemical properties of materials affect how hot they get when they absorb solar energy. Water heats up slowly and soil heats up rapidly when they absorb the same amount of energy. Water cools slowly and soil cools rapidly when they are moved to the shade.

Solar Energy in Action

Imagine a summer trip to the beach. On a cloudless day, the Sun shines down with equal intensity on the sandy beach and the ocean. It's a hot day.

When the car stops in the parking lot in the early afternoon, the pavement in the parking lot is hot. The black asphalt has absorbed a lot of solar energy, and its temperature is 50 degrees Celsius (°C). You walk across the parking lot (ouch, hot!) and onto the white sand. Whew! The white sand isn't as hot. It is a bearable 32°C. You keep moving toward the water. You finally get relief from the intense heat. The temperature of the ocean water is 22°C.

The asphalt, sand, and ocean water were all exposed to the same intensity of solar energy for the same length of time. But they are all different temperatures.

Black asphalt absorbs a lot of energy and gets very hot. White sand reflects a lot of solar energy. Light-colored sand doesn't get as hot as asphalt. Water absorbs a lot of energy, but it stays cool.

The temperature of Earth's surface is not the same everywhere. Land gets hotter than water in the sunlight. Land gets colder than water when the Sun goes down. Land heats up and cools off rapidly. Water heats up and cools off slowly. The result is uneven heating of Earth's surface.

Uneven Heating Worldwide

You can feel uneven heating of Earth when you are barefoot on asphalt.

You can experience uneven heating of Earth's surface with your bare feet when walking on asphalt. The difference in temperature between the asphalt and water is obvious. On a larger scale, the whole planet is heated unevenly. The tropical areas (the tropics, near the equator) are warmer. The polar areas are cooler. That's because the angle of exposure of the solar energy is more direct in the tropics.

The illustration shows how sunlight comes straight down on the tropics. But the sunlight comes at a sideways angle toward the poles. You can see how the same amount of light is spread over a much larger area in the north than in the tropics. This results in more heating in the tropics and less heating in the northern areas.

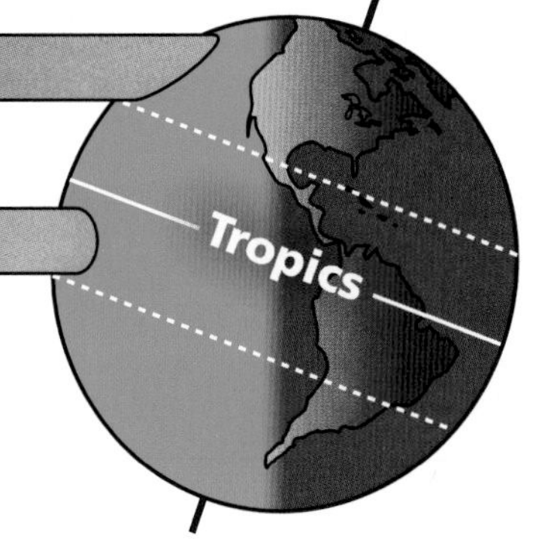

Two identical beams of sunlight. The upper beam spreads over a larger area toward the poles.

Thinking about Uneven Heating

1. What causes Earth's surface to heat up?
2. What are some of the variables that cause uneven heating of Earth's surface?
3. What happens to the temperature of equal volumes of soil and water when they are placed in the sunlight for 30 minutes?

Heating the Air: Radiation and Conduction

You might have had an experience like this one. The campfire has burned down to a bed of hot coals. Now it is time to toast some marshmallows. You put a marshmallow on a long stick and stand at a safe distance from the coals to toast your treat. You can feel the heat coming from the coals. After a minute, the marshmallow is brown and gooey. You pop it into your mouth. Yikes, that's hot! You didn't wait long enough for it to cool.

That story includes a couple of heat experiences. Have you ever stopped to think about what heat really is? What is the heat that you felt coming off the coals and the heat in the marshmallow that burned your tongue?

Heat = Movement

Objects in motion have energy. The faster they move, the more energy they have. Energy of motion is called **kinetic energy**.

Matter, like soccer balls, juice bottles, water, and air, is made of particles that are too small to see. The particles are in motion. They are in motion even in steel nails and glass bottles. In solids, the particles vibrate back and forth. In liquids and gases, the particles move all over the place. The faster the particles vibrate or move, the more energy they have.

Particle motion is kinetic energy, which can produce heat. The amount of kinetic energy in the particles of a material determines how much heat it produces. The particles in hot materials are moving fast. The particles in cold materials are moving more slowly.

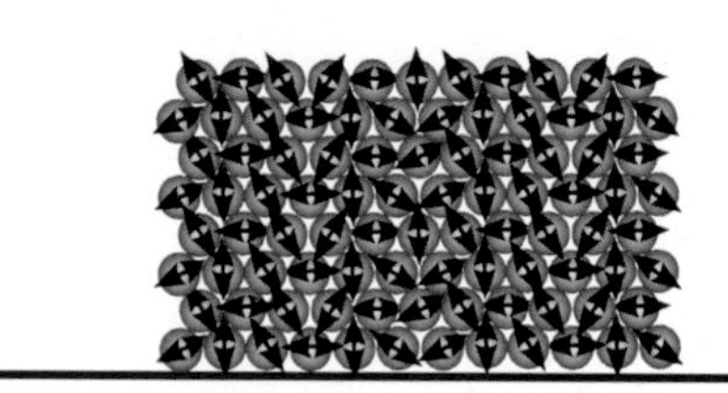

The particles in solids are held close to each other. They move by vibrating.

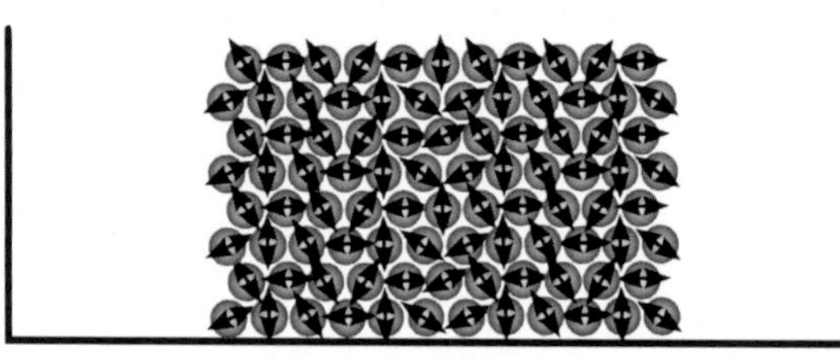

When solids get hot, the particles vibrate more. Hotter solids have more kinetic energy than colder solids.

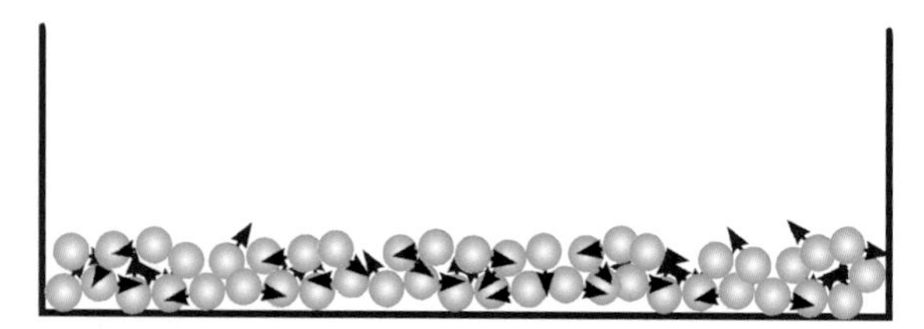

The particles in liquids move by bumping and sliding around each other.

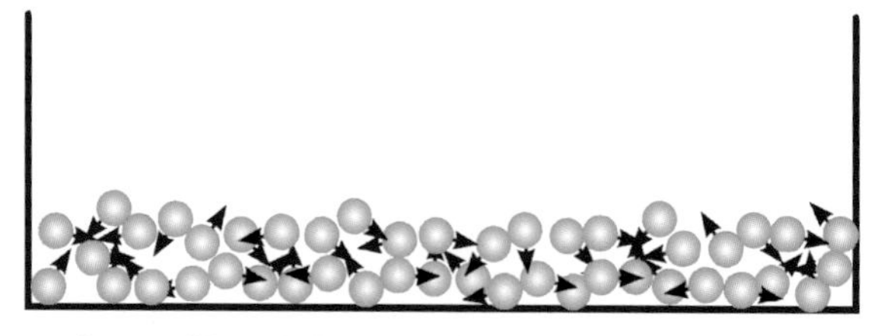

When liquids get hot, the particles bump and push each other more. Hotter liquids have more kinetic energy than colder liquids.

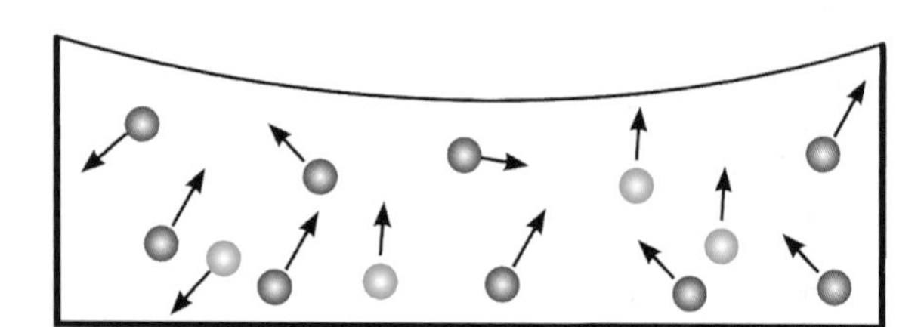

The particles in gases fly individually through the air in all directions.

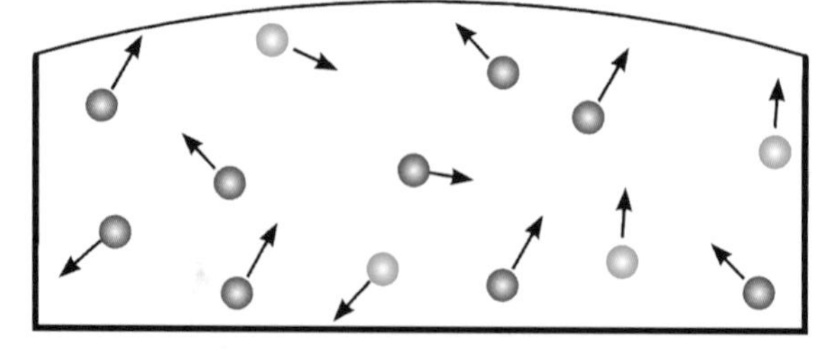

When gases get hot, the particles fly faster and are farther apart. Hotter gases have more kinetic energy than colder gases.

Energy Transfer

Heat can transfer (move) from one place to another. We can observe the **energy transfer** when energy is present as heat. Scientists sometimes describe heat transfer as heat flow, as though it were a liquid. Heat is not a liquid, but flow is a good way to imagine its movement.

Heat flows from a hotter location (high energy) to a cooler one (less energy). For example, if you add cold milk to your hot chocolate, heat flows from the hot chocolate to the cold milk. The hot chocolate cools because heat flows away. The cold milk warms because heat flows in. Soon the chocolate and the milk arrive at the same warm temperature, and you gulp them down.

Energy Transfer by Radiation

Burning gas, like the burner on a stove, can get very hot. When this happens, the burner is radiating heat and light. If you are close to a lightbulb, you can see light and feel heat, even though you are not touching the lightbulb. Energy that travels through air is **radiant energy**.

Radiant energy travels in rays. Heat and light rays radiate from sources like hot campfire coals, lightbulbs, and the Sun.

Radiation from the Sun passes through Earth's atmosphere. We call this solar energy. When solar energy hits a particle of matter, such as a gas particle in the air, a water particle in the ocean, or a particle of soil, the energy can be absorbed. Absorbed radiation increases the kinetic energy (movement) of the particles in the air, water, or soil. Increased kinetic energy produces a higher temperature, so the material gets hotter.

Radiation is one way energy moves from one place to another. Materials don't have to touch for energy to transfer from one place to the other.

Energy Transfer by Conduction

Imagine that hot toasted marshmallow or maybe a slice of pizza straight from the oven going into your mouth. This is another kind of energy transfer. When energy transfers from one place to another by contact, it is called **conduction**.

The fast-moving particles of the hot pizza bang into the slower particles of your mouth. The particles in your tongue gain kinetic energy. At the same time, particles of the hot pizza lose kinetic energy, so the pizza cools off. Some of the pizza's kinetic energy is conducted to heat receptors on your tongue, which sends a message to your brain that says, "Hot, hot, hot!"

When you heat water in a pot, the water gets hot because it touches the hot metal of the pot. Kinetic energy transfers from the hot metal particles to the cold water particles by contact, which is conduction.

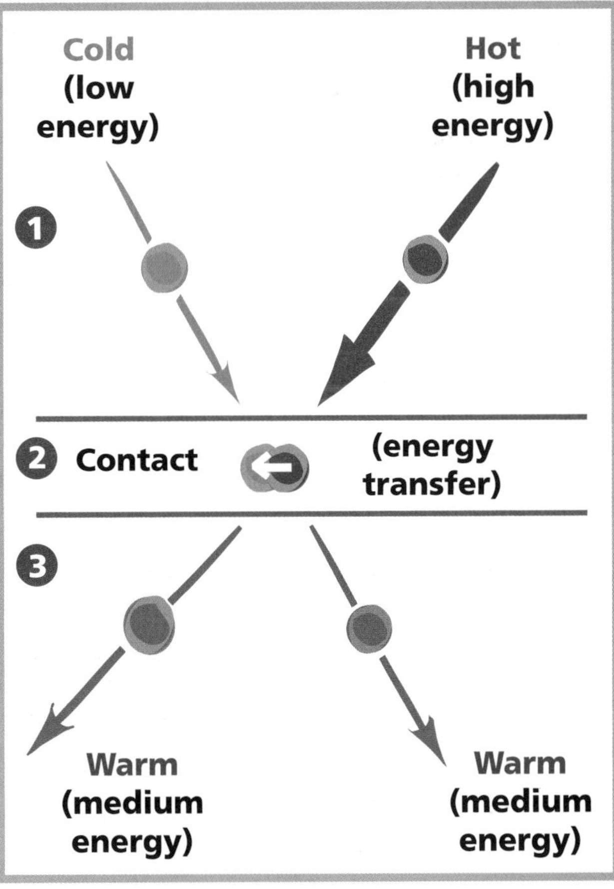

A "hot" particle with a lot of kinetic energy collides with a "cold" particle with little kinetic energy. Energy transfers at the point of contact. The cold particle gains energy, and the hot particle loses energy.

Energy Transfer to the Air

We have learned about energy transfer by radiation and conduction. Now let's explore energy from the Sun and what happens when it interacts with the air. Many kinds of rays radiate from the Sun. The most important rays are visible light and infrared light, which is invisible. So how do the Sun's rays heat up the air?

Air is 99 percent nitrogen and oxygen particles. Neither kind of particle absorbs visible light or infrared radiation. Only water vapor and carbon dioxide absorb significant amounts of radiant energy, and this is mostly infrared rays, not visible light.

If only a tiny part of the air absorbs the incoming radiant energy, how does the rest of the air get hot? We need to consider parts of two other systems, the **hydrosphere** (water) and **geosphere** (land).

Earth's surface absorbs visible light. The land and water warm up. The air particles that touch the warm land and water particles gain energy by conduction. But there is more.

The warm land and water also radiate energy. This is a very important idea. Earth gives off infrared radiation that can be absorbed by water particles and carbon dioxide particles in the air. The energy transferred to the small number of water particles and carbon dioxide particles is transferred throughout the air by conduction. This happens when energized water particles and carbon dioxide particles bang into oxygen and nitrogen particles.

The air is not only heated from above. It is also heated from below.

Wind and Convection

Kite flying can be a lot of fun if the conditions are right. If the conditions are wrong, kite flying can be boring. What makes conditions right for kite flying? Wind.

Wind lifts a kite high in the air.

Wind is air in motion. Air is matter. Air has mass and takes up space. When a mass of air is in motion, it can move things around. Wind can blow leaves down the street, lift your hat off your head, and carry a kite high in the air.

Sometimes air is still. Other times the wind is blowing. What causes the wind to blow? What puts the air into motion? The answer is energy. It takes energy to move air. The energy to create wind comes from the Sun.

Air is particles of nitrogen, oxygen, and a few other gases. The particles are flying around and banging into each other, the land, and the ocean. Let's imagine we are at the beach. It's early morning. The air over the land and the air over the ocean are both the same cool temperature.

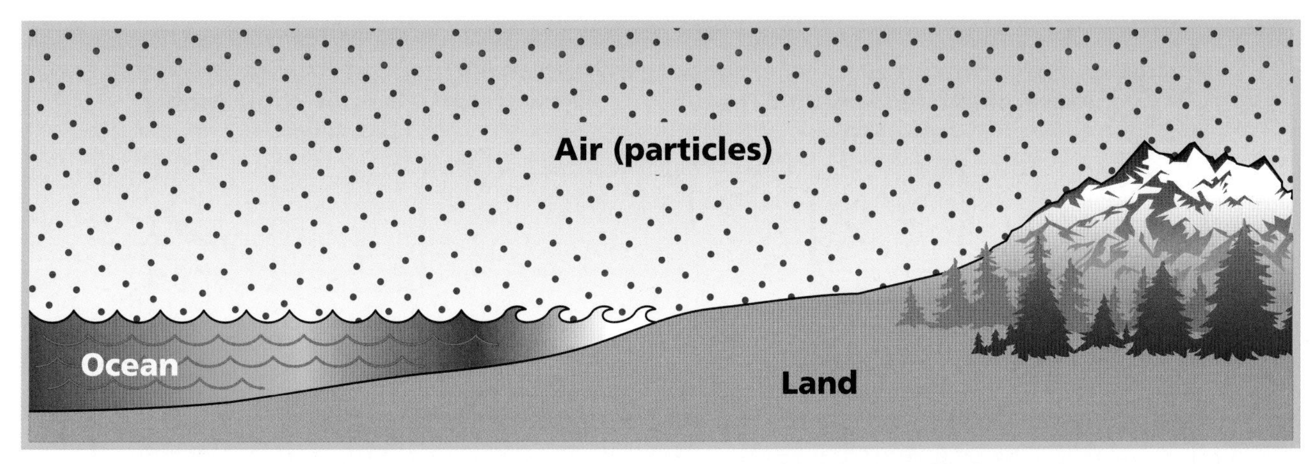

In the early morning, the land, ocean, and air are all the same cool temperature.

As the Sun shines down on the land (part of the geosphere) and ocean (part of the hydrosphere), solar energy is absorbed. The land heats up quickly. The ocean heats up very slowly. By noon the land is hot, but the ocean is still cool. Earth's surface is heated unevenly. The afternoon wind starts. Here's why.

When air particles bang into the hot surface of the land, energy transfers to the air particles. Because of this energy transfer, the air particles fly around faster. The air gets hot. The hot-air particles bang into each other harder. That pushes the particles farther apart.

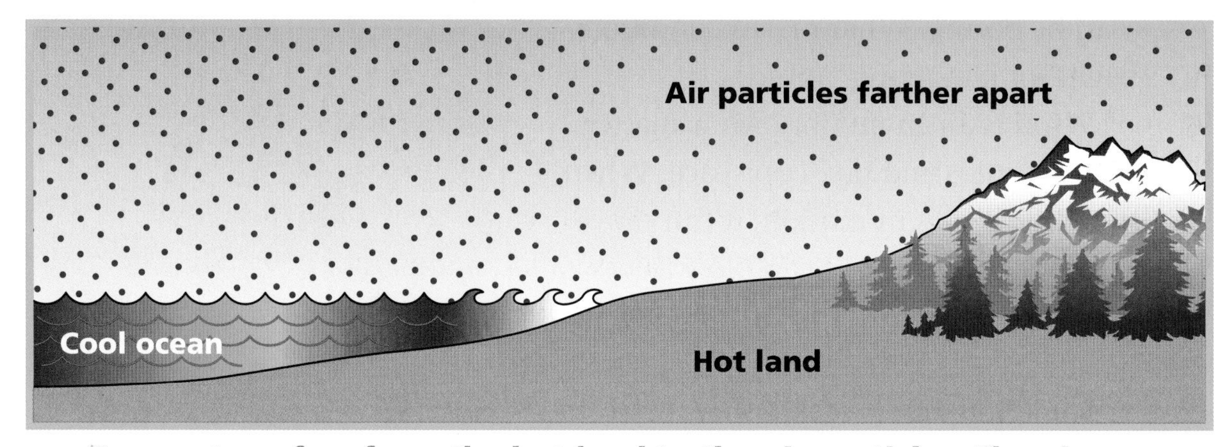

Energy transfers from the hot land to the air particles. The air particles move farther apart.

Over the ocean, air particles are banging into the cool surface of the water. The air stays cool. The air particles continue to move at a slower speed. The cool-air particles don't hit each other as hard, so they stay closer together.

A cubic meter of hot air has fewer particles than a cubic meter of cool air. The hot air is less dense than an equal volume of cool air.

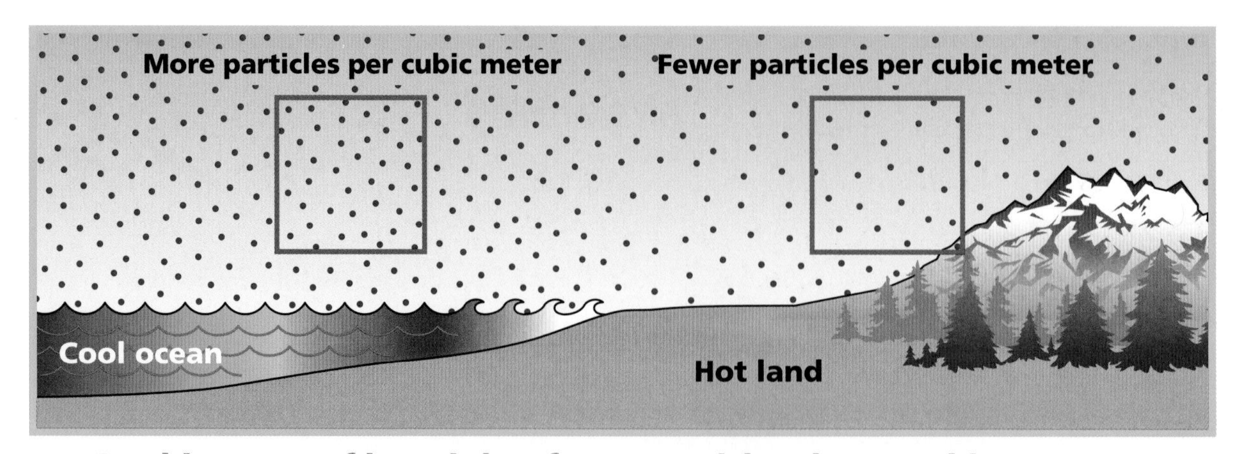

A cubic meter of hot air has fewer particles than a cubic meter of cold air. Hot air is less dense than cool air.

The Wind Starts

You know that cork floats on water. Cork floats on water because it is less dense than water. If you take a cork to the bottom of the ocean and let it go, it will float to the surface.

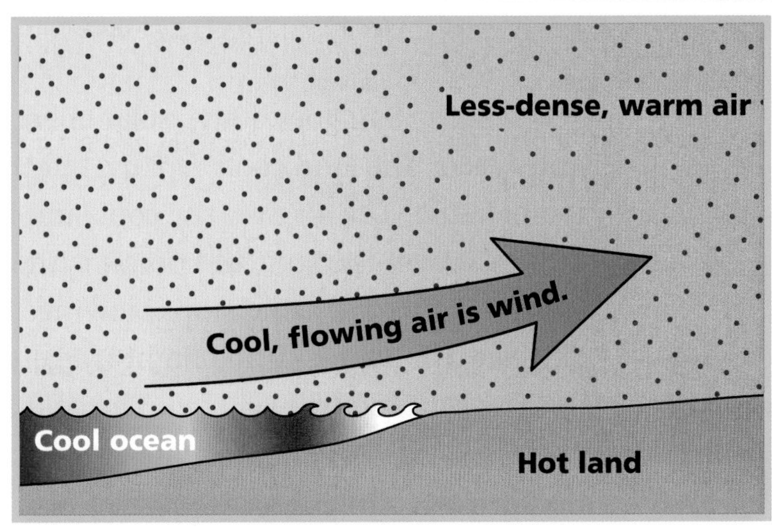

Dense, cool air flows from the ocean to the land. Less-dense, warm air rises.

That's exactly what happens with warm and cold air. The warm air over the land floats upward because it is less dense than the cool air over the ocean. The more-dense, cool air flows into the area where the less-dense, warm air is and pushes it upward. The movement of more-dense air from the ocean to the warm land is wind. Wind is the movement of more-dense air to an area where the air is less dense.

There is more to the story of wind. Two things happen at the same time to create wind. The warm air cools as it rises, becoming more dense than the surrounding air. At the same time, the dense air from the ocean warms up as it flows over the hot land.

As a result, air starts to move in a big circle. Air that is warmed by the hot land moves upward. The warm air cools as it moves up, gets more dense, and starts to fall. The rising and falling air sets up a big circular air current. The circular current is called a **convection current**.

As long as Earth's surface continues to be heated unevenly, the convection current will continue to flow. The part of the convection current that flows across Earth's surface is what we experience as wind. But what happens at night?

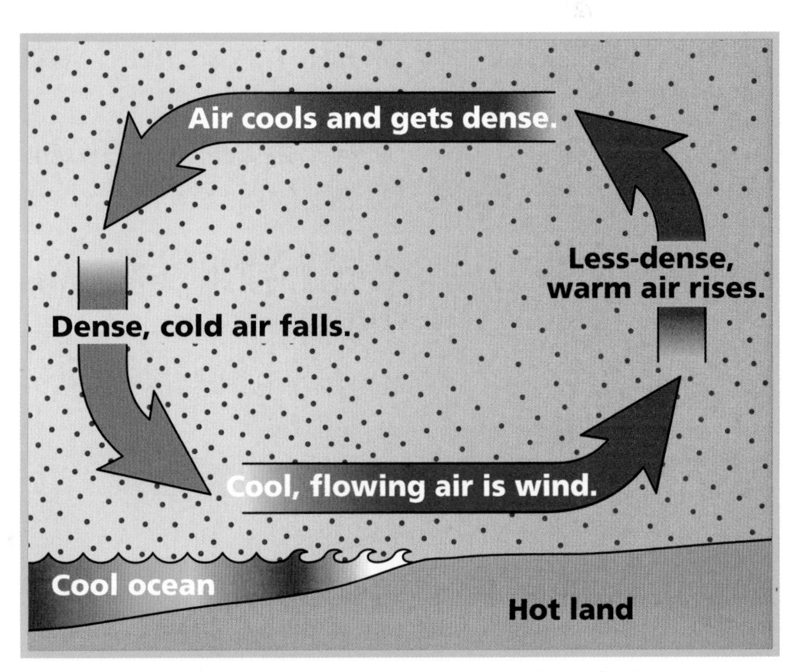

A convection current is the result of the uneven heating of Earth's surface.

The Wind Changes Direction

When the Sun goes down, solar energy no longer falls on the land and ocean. The land cools rapidly, but the ocean stays at about the same temperature. The air over the cool land is no longer heated. The density of the air over the land and ocean is the same. The convection current stops flowing. The wind stops blowing.

What will happen if the night is really cold? The land will get cold. The air over the land will get cold. The cold air will become more dense than the air over the ocean. The more-dense air will flow from the land to the ocean. The convection current will flow in the opposite direction, and the wind will blow from the land to the ocean.

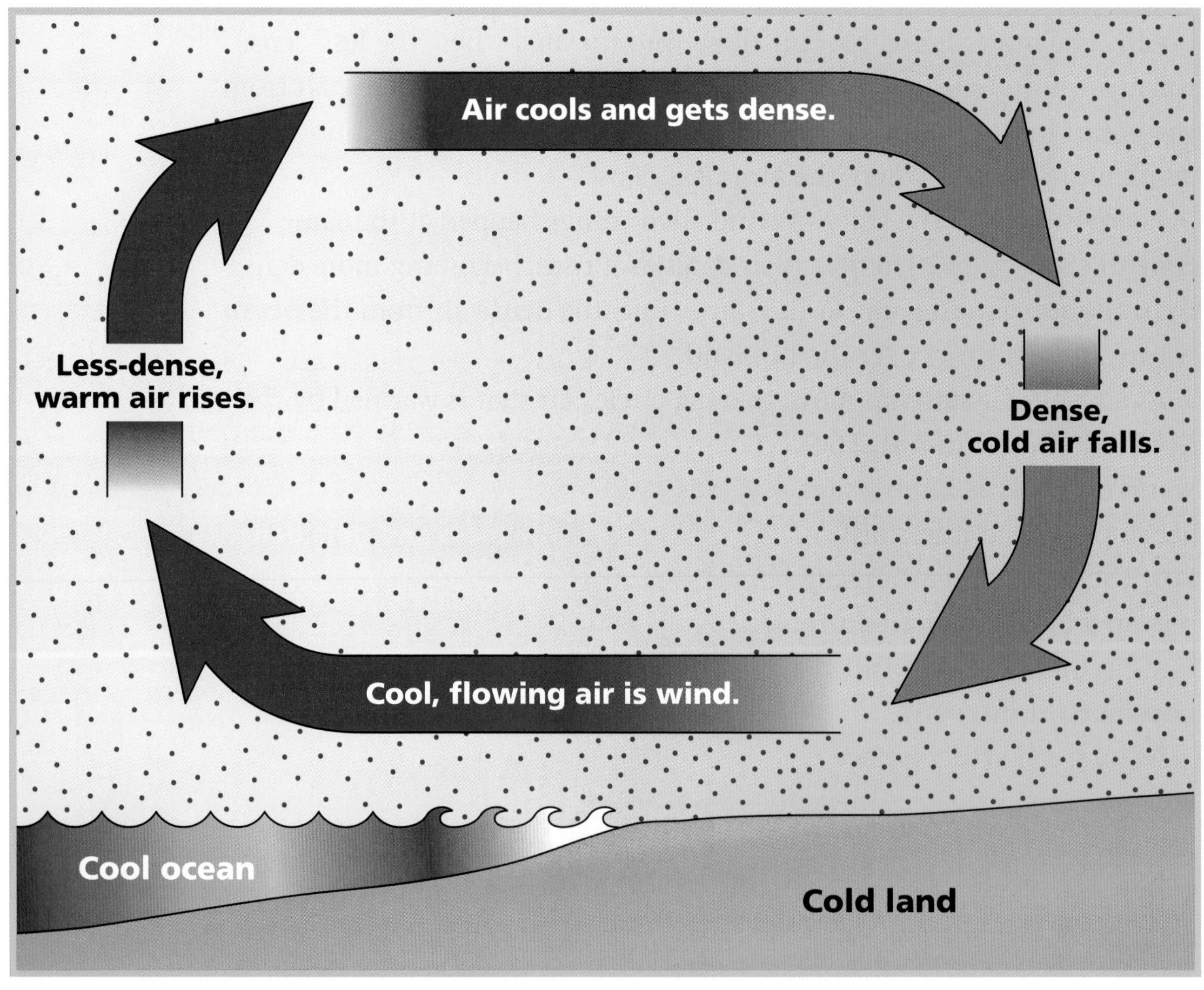

This convection current creates wind that blows from the land to the ocean.

Convection Summary

Uneven heating of Earth's surface by the Sun causes uneven heating of the air over Earth's surface. Uneven heating of air causes wind. Warm air is less dense than cold air. Cold, dense air flows to an area where the air is warmer and less dense. The less-dense air is pushed upward. As the warm air moves upward, it cools. Cool air is more dense, so it falls back to Earth. This circular pattern of air flow is a convection current. Convection currents are important ways that air masses move from place to place in the atmosphere. Convection currents transport energy from place to place.

Convection currents produce wind. The greater the difference in temperature between the warm and cold air masses, the harder and faster the wind will blow. Uneven heating of Earth's surface is the cause of many weather changes on Earth, including hurricanes, **tornadoes**, and **thunderstorms**.

Convection currents produce wind.

Uneven heating of Earth can cause hurricanes.

Thinking about Wind

1. How are convection currents produced in the air?
2. Explain what causes wind.
3. What happens to air particles when air is heated?
4. What is the source of energy that causes the wind to blow?

Wind Power

Heat isn't the only energy resource that starts with the Sun. Wind, which is caused by the Sun's uneven heating of Earth, can also be used to generate power.

Wind energy has powered sailing ships for thousands of years.

Wind is created when cool air rushes in to take the place of warmer, less-dense air. People have used wind power for thousands of years. In the past, sailing ships were the quickest and easiest way to travel. These ships had large sails to catch the wind and move them across the water.

One ancient use of wind power is windmills. Arabic people introduced windmills to Europe in about the 12th century. Ancient windmills were used to grind grain into flour. These windmills caught the wind in sails made of wood and cloth. The sails turned an axle, which transferred its motion to a turning pole. The pole turned the millstones that ground the grain. Windmills were also used to pump water.

Inventors and scientists have made many changes in windmills. Modern windmills are called wind turbines. A small motor starts the blades turning. Then they can turn on their own, even in wind speeds as low as 24 km per hour.

Windmills have provided power for pumping water and grinding grain in Europe since the 12th century.

Wind turbines convert wind energy into electricity.

Sometimes dozens or even hundreds of wind turbines are set up on a wind farm. One well-known wind farm is at Altamont Pass in California. As warm valley air rises in the central part of the state, cool air from the ocean flows over the coastal mountains. Narrow gaps between mountains, called passes, channel the air and increase its speed.

There are over 13,000 wind turbines in Altamont Pass and two other areas in California. In 1995, these three areas produced 30 percent of the world's wind-generated electricity. Roscoe Wind Farm in Texas is the largest land-based wind farm in the world. It has 627 wind turbines over 100,000 acres of west Texas. This wind farm can provide electricity to 230,000 homes. In 2010, wind power provided about 2 percent of the electricity generated in the United States.

Wind power has some disadvantages. The power produced depends on how strongly the wind is blowing. Therefore, wind farms can produce too much energy at some times and too little energy at others. Excess energy needs to be stored. It can be used to heat water or oil, or stored in batteries.

Some people don't like wind farms because they think the turbines are ugly. The turbines also create a whipping sound as they turn. The sound annoys some people and scares away animals. Birds can be killed by flying into spinning turbines.

Even though wind power has its problems, it is still an important source of energy. Each year, the United States uses the same amount of energy from wind turbines as 716,400 barrels of oil could provide. Using wind power instead of burning **fossil fuels** keeps about 2.5 million tons of carbon out of the air. Eventually fossil fuels will start to run out, and traditional methods of creating energy will become less desirable. Wind power, like solar power, is an energy source of the future.

Solar Technology

People use solar energy in a number of ways. People use it to heat water, to cook food, to warm and cool buildings, and to generate electricity.

Solar Water Heaters

Solar water heaters have been around for thousands of years. The ancient Romans created public baths supplied with water that flowed through heating channels. These channels were like canals, open to the Sun. The water channels were lined with black slate to absorb as much of the Sun's energy as possible. Centuries later, people painted metal tanks black and tilted them toward the Sun to warm the water inside.

In 1891, the first American commercial solar water heater, the Climax, was built. It was a black iron tank inside a wooden box. The box was lined with black felt and covered with glass. It sat on a roof exposed to the Sun.

The Climax worked fine on a sunny day. But what about cloudy days or nighttime? In 1909, the Day and Night water heating system was invented. This system included a solar heater and a separate insulated storage tank. After the Sun heated the water, it was piped into the storage tank where it remained warm.

Today, there are many types of solar water heaters, some of them shown on these pages.

A thermosyphon system solar water heater

A flat-plate collector solar water heater

One type of solar water heater available today uses a flat-plate solar collector in an insulated box. On top of the collector are small tubes filled with water to absorb the Sun's rays. The system is placed on the roof of a house. The Sun-heated water is piped down into an insulated storage tank in the house.

The many types of modern solar water heaters share some common features. They must be positioned to capture the Sun's energy so they are often on top of roofs. They are made of materials that absorb the heat energy and transfer the energy to water. The systems are insulated so the heat energy doesn't escape from the water. The water moves through pipes to a storage tank and is accessible by turning a faucet.

A glass tube collector solar water heater

Solar Cookers

Most of your meals are probably cooked in an oven or on a stove powered by electricity or natural gas. Electric microwave ovens are also popular for cooking food. But what about cooking where there is no access to gas or electricity? In developing countries, more than 2 billion people burn wood for cooking. But burning wood is bad for the environment and creates health risks for people. Large numbers of trees have to be cut down. Wood burning creates smoke that pollutes the air and causes breathing problems.

People can use solar cookers as a safe, inexpensive, pollution-free way to cook. A number of organizations have developed solar ovens in countries all over the world.

A solar cooker in the Zanskar Mountains of northern India

One of the most successful solar cooking projects is in Kenya. Professor Daniel Kammen, of the University of California, Berkeley, has introduced solar cookers to villages in Kenya. His simple solar ovens are boxes lined with reflective foil and covered with glass. First, he demonstrates how easy they are to use. Kammen and his coworkers mix up a stew or cake. They put it in the oven. Then Kammen explains how the oven works. He also explains how to build one. Finally he opens the oven and lets everyone eat some of the cooked food.

A solar oven in Kenya

It takes about 3 hours to cook a pot of stew in a solar oven and about half an hour to boil a pot of water. Cooking food over a traditional fire takes less time. However, solar ovens eliminate the time-consuming chore of gathering firewood. Solar cooking also eliminates many respiratory illnesses, which are a common cause of death in Kenya.

If the community is interested in the solar oven, Kammen and his coworkers come back with plywood, foil, glass, and nails. They teach the villagers how to build their own ovens. So far, Kammen and his team have introduced hundreds of solar ovens in eastern Kenya.

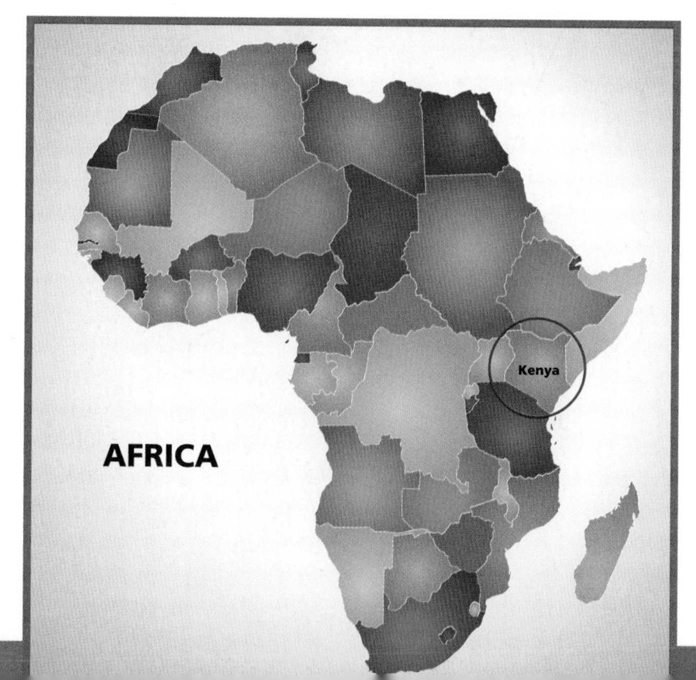

Solar Electricity

The Sun's energy can be converted into electricity. This process uses solar cells or steam-powered turbines. Solar cells are devices used to transform solar energy directly into electricity.

Solar cells are made of silicon, which is found in sand. The silicon is purified, crystallized, and cut into wafers or squares. Two wafers or squares are sandwiched together to form a solar cell. The cells are arranged on a flat panel made of glass, metal, or wood and sealed behind glass or plastic. One solar cell with a diameter of 10 centimeters (cm) can produce 1 watt of electric power. The electricity is then used immediately or stored in batteries.

Many solar cells are connected to make large panels called solar arrays. Solar arrays can generate enough electricity to store. The biggest challenge facing the use of solar cells is to bring down their cost. If this happens, many energy experts believe that solar cells could provide most of our energy needs by the end of this century.

Solar furnaces use huge mirrors to concentrate a lot of light in one place. Large solar furnaces can use this energy to produce steam, which can be used to turn large generators to produce large quantities of electricity. The solar furnace in Odeillo, France, has an 8-story-tall mirror and can produce temperatures up to 3,000 degrees Celsius (°C). That's hot enough to melt a steel plate in just 3.5 seconds!

The solar furnace in Odeillo, France

Maria Telkes

Dr. Maria Telkes

Dr. Maria Telkes (1900–1995) was the world's most famous female inventor in the field of solar energy. Some people even called her the Sun Queen!

Maria Telkes was born in Hungary. She lived most of her life in the United States. She first became interested in solar energy when she was with the Massachusetts Institute of Technology (MIT) in the 1940s. During her 50-year career, she worked for many top universities.

Solar cooking was of special interest to Telkes. She realized that we need to **conserve** fuel. The heat of the Sun provided a clean source of energy. During the 1950s, she invented a model of a solar oven that is still used today. Later the Ford Foundation gave Telkes a $45,000 grant to work on her solar oven. This allowed Telkes to improve her design.

Telkes also helped to create a solar house. In 1981, she worked to design and build the Carlisle House in Carlisle, Massachusetts. MIT and the US Department of Energy also worked on the project. This house features a solar heating and cooling system. It uses no fossil fuels, such as natural gas, oil, and coal. The house generates so much power that it is able to share its extra energy with the local utility company.

The solar-powered Carlisle House was built by Telkes in 1981.

Solar Buildings

Imagine a house or office building where the electric, heating, and cooling systems don't rely on oil, natural gas, or coal. Instead, all these systems use energy from the Sun for power. Homes like this do exist. They are called solar houses.

Solar houses have many different systems that use energy from the Sun. Some of these systems are passive. They don't use any mechanical devices such as pumps or generators. An example of passive space heating would use large windows that face the Sun. The windows let sunlight shine in to warm the house. When the weather is warm, window coverings come down over the windows and keep the house cool. Solar houses are designed to work with the Sun to maintain a comfortable temperature at all times of the year.

Other solar systems are active. They use mechanical devices, such as fans and pumps, to move captured solar energy throughout the house. For example, many solar houses use flat-plate collectors to heat water and air. A flat-plate collector sits in a box that is insulated on the bottom and sides. The top is covered by one or more layers of clear glass or plastic. The system is placed on the roof of a house. The visible light goes through the glass, where it is absorbed by the plate and converted into heat. The glass traps the heat inside the box. The hot air is pumped to rooms in the house where it heats the space.

Solar energy is good for the environment! That's why some houses and apartment buildings use solar panels.

Solar energy can also be used to cool buildings during the summer. Most solar air-conditioning systems use solar collectors and materials called desiccants. Desiccants can absorb large amounts of water. Fans force air from outdoors through the desiccants, which remove moisture from the air. Next the dry air flows through a heat exchanger that removes some of the heat. Then the air passes over a surface soaked with water. As the water comes in contact with the dry air, it evaporates and removes more heat from the air. Finally the cooled air is pumped throughout the building.

Other types of active systems involve solar cells. Solar arrays (also called solar panels) can generate electricity for all kinds of household uses. Solar panels can be placed on the roof of a house or on the sides of apartment buildings. It is important that the solar cells capture as much sunlight during the day as possible.

Solar houses have many advantages. They are environmentally friendly. Solar energy is a **renewable resource**. It does not use up Earth's resources the way burning fossil fuels or wood does. In addition, solar collectors are pollution-free. They don't create fumes or other dangerous chemicals that can poison the environment or make people sick.

Thinking about Solar Technology

1. What are some features of solar water heaters used today?
2. What are the advantages of solar cookers?
3. What did Maria Telkes contribute to solar technology, and when did she do this?

Condensation

When water evaporates, where does it go? It goes into the air. Water is always evaporating. Clothes are drying on clotheslines. Wet streets are drying after a rain. Water is evaporating from lakes and the ocean all the time. Every day more than 1,000 cubic kilometers (km) of water evaporates worldwide. And all that water vapor goes into the air! That amount of water would cover the entire state of California 3 meters (m) deep.

What happens to all that water in the air? As long as the air stays warm, the water stays in the air as water vapor. Warmth (heat) indicates the presence of energy. As long as the particles of water vapor have a lot of kinetic energy, they continue to exist as gas.

But if the air cools, things change. As the air cools, the particles lose kinetic energy and slow down. When this happens, particles of water vapor start to come together. Slowing down and coming together is called condensation. Condensation is the change from gas to liquid.

Particles of condensed water vapor form tiny masses (droplets) of liquid water. When invisible water vapor in the atmosphere condenses, the water becomes visible again. Clouds and **fog** are made of these tiny droplets of liquid water.

The clouds in the sky are made of tiny droplets of water.

Condensation usually happens on a cold surface. In class, you observed condensation on the cold surface of a plastic cup filled with ice water. But there are no cups of ice water in the sky. What kind of surface does water vapor condense on?

Most condensation in the air starts with dust particles. Water particles attach to a dust particle. When a tiny mass of water has attached to a dust particle, other water particles will join the liquid mass.

When you look up in the sky and see clouds, you are seeing trillions of droplets of liquid water. Each droplet is made up of billions of water particles, but a single droplet is still too small to see. You can see them when trillions and trillions of them are close together in clouds.

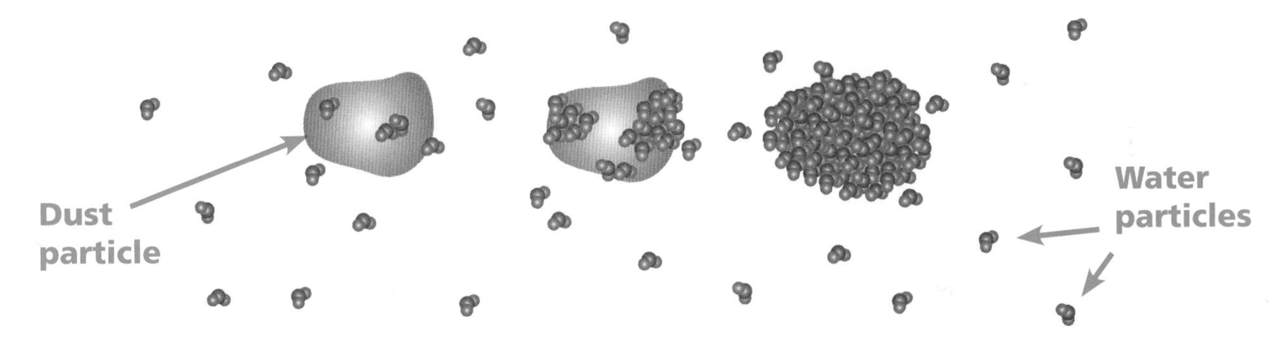

The mass of water grows and grows until it forms a tiny droplet of water.

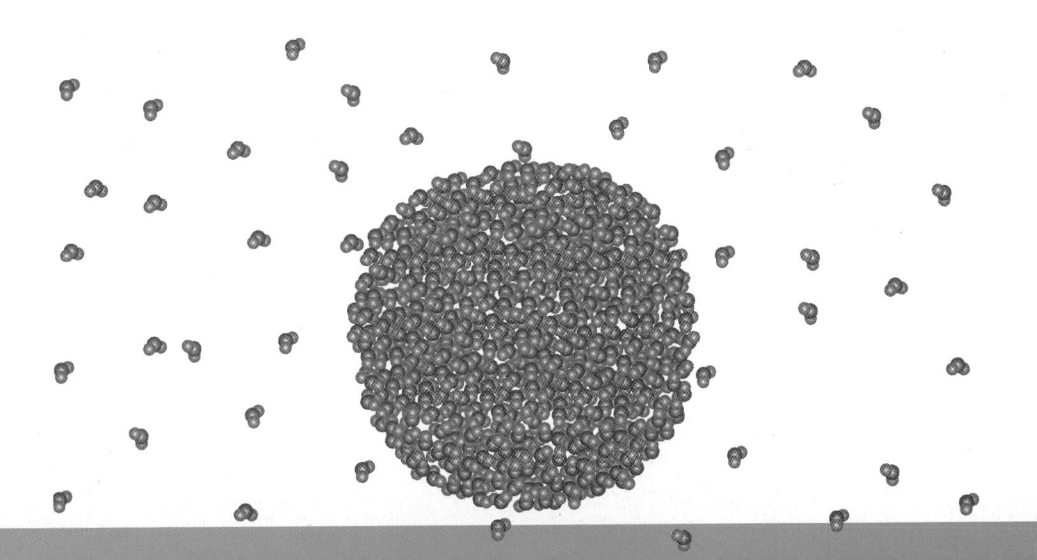

Fog close to the ground

Dew on a spider's web

Dew on a flower

Where else have you seen condensation other than up in the clouds? Sometimes water vapor condenses close to the ground. This is called fog. Being in fog is really being in a cloud that is at ground level.

Water vapor doesn't always condense in air. If you go out early in the morning after a warm day, you might see condensation called **dew**. In these pictures, dew formed on a spider's web and on a flower.

Water vapor condenses indoors, too. On a cold morning you might see condensation on your kitchen window. Or if you go outside into the cold wearing your glasses, they could get fogged with condensation when you go back inside.

Condensation on a window

What happens to the bathroom mirror after you take a shower? The air in the bathroom is warm and full of water vapor. When the air makes contact with the cool mirror, the water vapor condenses on the smooth surface. That's why the mirror is foggy and wet.

When the temperature drops below the freezing point of water (0°C), water vapor will condense and freeze. Frozen condensation is called **frost**. Frost is tiny crystals of ice. Frost might form on a car window on a cold night. You can also see frost on plants early on a winter morning. But you have to get up before the Sun if you want to see the beautiful frost patterns.

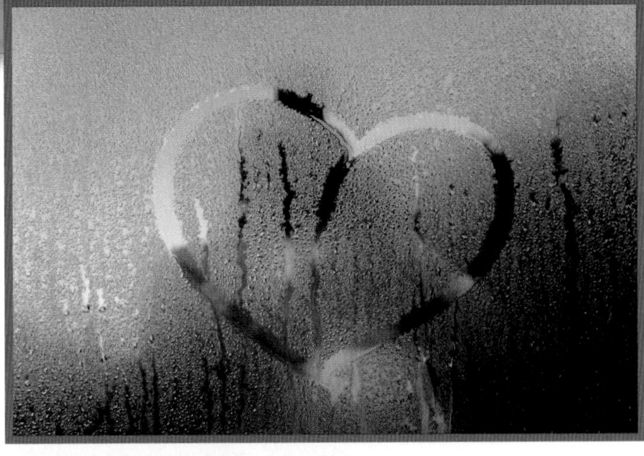

Condensation on a mirror

Frost on a window

Frost on grass and a leaf

Thinking about Condensation

1. What is condensation?
2. What role does temperature play in condensation?
3. What is frost?
4. Why does condensation form on a glass of iced tea?

Where Is Earth's Water?

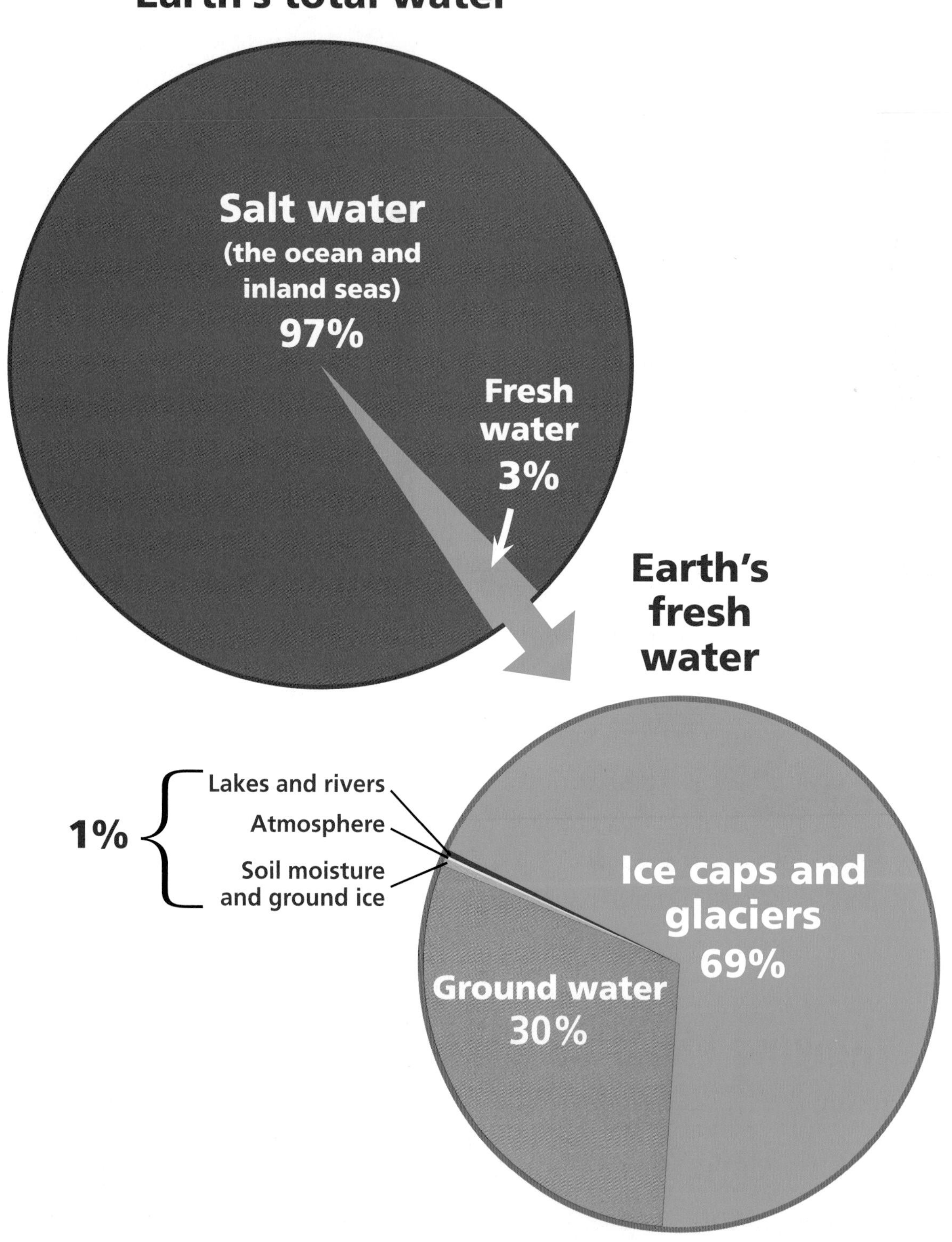

The Mississippi River

The Water Cycle

Water particles in the water you drink today might have once flowed down the Mississippi River. Those same particles might have washed one of George Washington's shirts. They might even have been in a puddle lapped up by a thirsty dinosaur a million years ago!

Water is in constant motion on Earth. You can see water in motion in rushing streams, falling raindrops, and blowing snowflakes. But water is in motion in other places, too. Water is flowing slowly through the soil. Water is drifting across the sky in clouds. Water is rising through the roots and stems of plants. Water is in motion all over the world.

Think about the Mississippi River for a moment. It flows all year long, year after year. Where does the water come from to keep the river flowing?

The water flowing in the river is renewed all the time. Rain and **snow** fall in and near the Mississippi River. Rain falling nearby soaks into the soil and runs into the river. The snow melts in the spring and supplies water for the river during the summer. Rain and snow keep the Mississippi River flowing.

The rain and snow in the Mississippi River are just a tiny part of a global system of water recycling. The system is called the **water cycle**.

The big idea of the water cycle is this. Water evaporates from Earth's surface and goes into the atmosphere. Water in the atmosphere moves to a new location. The water then returns to Earth's surface in the new location. The new location gets a new supply of water.

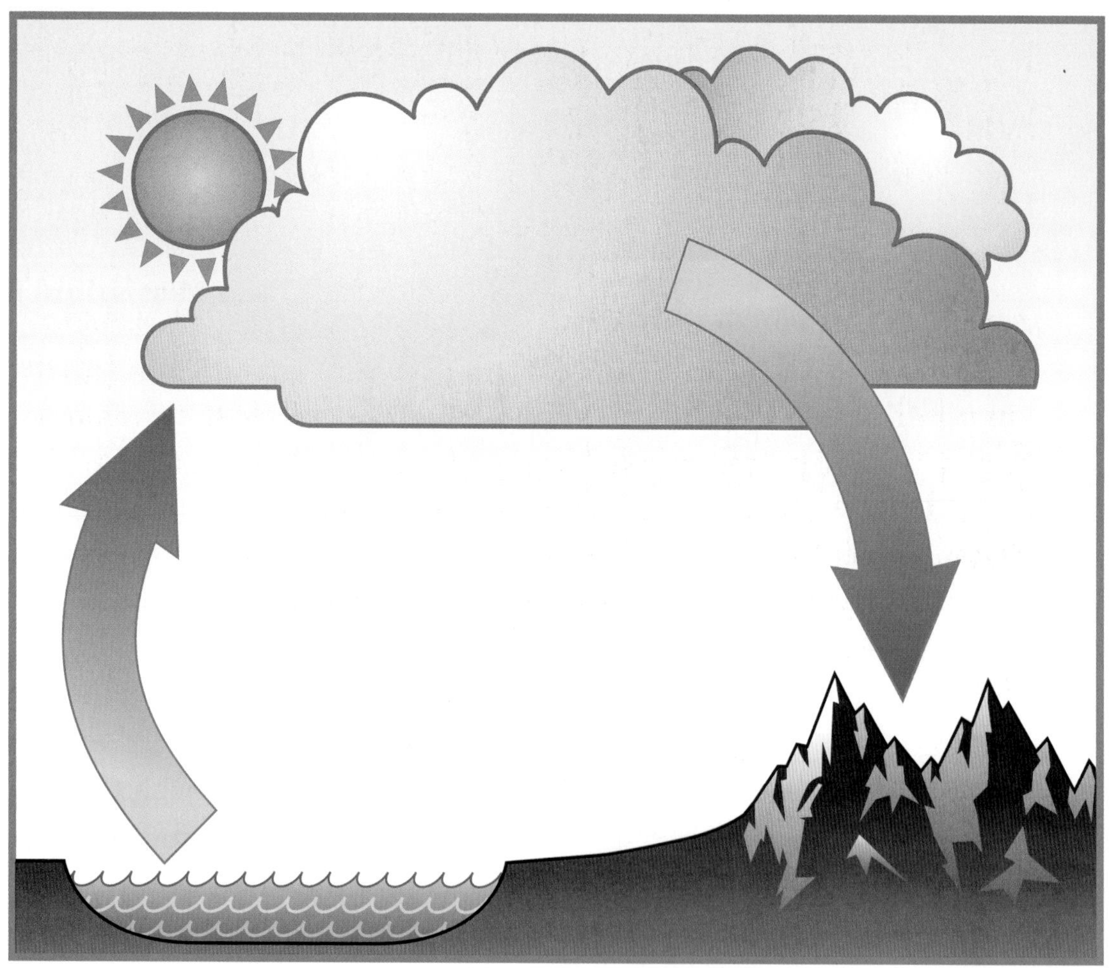

A simple water-cycle diagram

Water Evaporates from Earth's Surface

The Sun drives the water cycle. Energy from the Sun falls on Earth's surface and changes liquid water into water vapor. The ocean is where most of the evaporation takes place. But water evaporates from lakes, rivers, soil, wet city streets, plants, animals, and wherever there is water. Water evaporates from all parts of Earth's surface, both water and land.

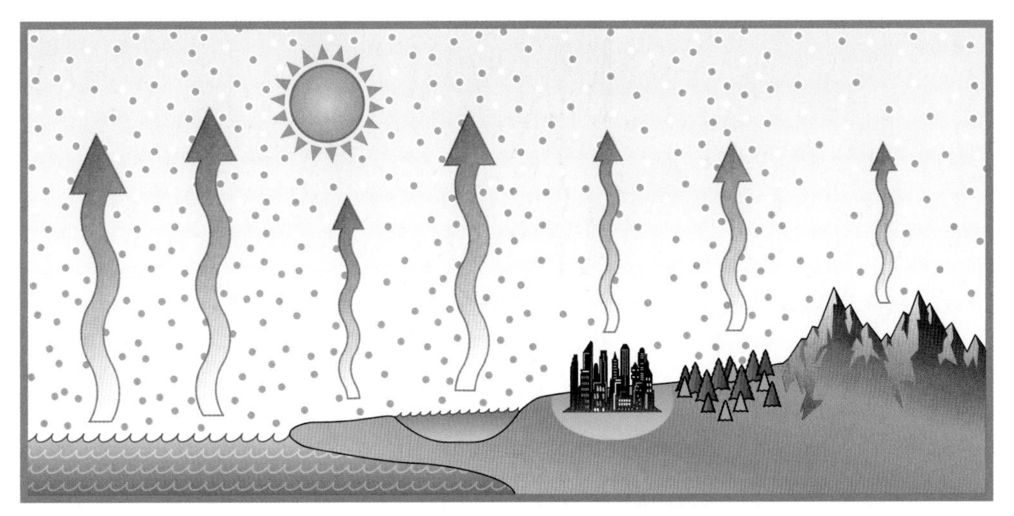

Water evaporates from all of Earth's surfaces.

Water vapor is made of individual water particles. Water vapor enters the air and makes it moist. The moist air moves up in the atmosphere. As moist air rises, it cools. When water vapor cools, it condenses. Water in the atmosphere changes from gas to liquid. Tiny droplets of liquid water form. The condensed water is visible as clouds, fog, and dew.

Water vapor condenses in the atmosphere to form clouds.

Water Falls Back to Earth's Surface

Wind blows clouds around. Clouds end up over mountains, forests, cities, deserts, and the ocean. When clouds are full of condensed water, the water falls back to Earth's surface as rain. If the temperature is really cold, the water will freeze and fall to Earth's surface as snow, **sleet**, or **hail**.

Water falls back to Earth's surface as rain, snow, sleet, or hail.

Water particles move through the water cycle at different speeds. They also follow different paths. For example, rain might soak into the soil. A particle might be taken in by plant roots. It might soon escape into the air through holes in plant leaves during a process called **transpiration**. If the air is cool, water might condense immediately as dew and fall back onto the soil. This is a very small water cycle that **recycles** water back to its starting place quickly.

Rain that lands on the roof of your school might flow to the ground. From there it could enter a stream. After a long journey, it could find its way to the ocean. There the rainwater could reenter the atmosphere as water vapor. By the time it condenses with millions of other particles to form a drop, the rainwater could be hundreds of kilometers (km) away from where it started. When the particle returns to Earth's surface, it could fall on the roof of a school in another state. This is an example of a large water cycle that moves water to a new location.

Rain can sink into the ground or freeze in a glacier. A particle far underground or deep in a mass of ice can take a long time to reenter the water cycle. It might take 100 years for a particle of ground water to come to the surface in a spring, and even longer for a particle to break free from a glacier.

The Sun provides the energy to change liquid water into vapor. Water vapor enters the air, where it is carried around the world. When water condenses, gravity pulls it back to Earth's surface. That's the water cycle, and it goes on endlessly.

Thinking about the Water Cycle

1. What is the water cycle?
2. When water falls from clouds, what forms can it take?
3. Describe a large water cycle that takes a long time to complete.
4. Describe a small water cycle that takes a short time to complete.

Severe Weather

Hurricane Katrina making landfall on the Gulf Coast

On August 29, 2005, Hurricane Katrina roared across the Gulf of Mexico and onto land. Throughout the country, people watched TV and listened to the radio as Katrina plowed into the states of Louisiana, Mississippi, and Alabama. The wind speed was 250 kilometers (km) per hour. The rain poured down. When the storm had passed, hundreds of people were dead, hundreds of thousands were homeless, and the city of New Orleans was flooded. The cost of the damage was in the billions of dollars.

Weather is fairly predictable most of the time. During the summer months in San Francisco, California, mornings and afternoons are often foggy. There might be sunshine in the middle of the day. In the winter months, rain is common. In Los Angeles, California, hot, dry weather is typical in the summer. In Gulf states, summer days are often hot and humid. In the Midwest and East, winters are usually cold, cloudy, and snowy. These are the normal weather conditions that people come to expect where they live.

It's the change from normal to the extreme that catches people's attention. Tornadoes, thunderstorms, windstorms, hurricanes, **droughts**, and floods are examples of **severe weather**. Severe weather brings out-of-the-ordinary conditions. It may cause dangerous situations that can damage property and threaten lives.

Rain is a common type of precipitation.

What Is Weather?

We are surrounded by air. It's a little bit like living on the bottom of an ocean of air. Things are always going on in the air surrounding us. The condition of the air around us is what we call weather.

A sunny day in Chicago, Illinois

Weather can be described in terms of four important variables. They are temperature, humidity, air pressure, and wind. They are called variables because they change. A day with nice weather might be warm, but not too hot. The sky is clear with just a little bit of moisture in the air. The air is still or moving with a light breeze. That's a perfect day for most people. But not too many days are perfect. Usually it's too hot, too humid, too windy, or too something. But don't worry. Weather always changes.

What Causes Weather to Change?

Energy makes weather happen. Energy makes weather change. The source of energy to create and change weather is the Sun.

When sunlight is intense, the air gets hot. When sunlight is blocked by clouds, or when the Sun goes down, the air cools off.

Moisture in the air takes the form of humidity, clouds, and precipitation. Intense sunlight evaporates more water from the land and ocean of Earth's surface. The result is more humidity, more cloud formation, and more rain. When sunlight is less intense, evaporation slows down.

Movement of air is wind. Uneven heating of Earth's surface results in uneven heating of the air touching Earth's surface. Warm air expands and gets less dense. More-dense, cool air flows under the warm air. This starts a convection current. The air flowing from the cool surface to the warm surface is wind.

When air pressure falls, rain is likely. A storm is possible.

Stormy weather approaching

Hurricane Earl near the Caribbean Islands in 2010

Hurricanes and Tropical Storms

Hurricanes are wind systems that rotate around an eye, or center of low air pressure. Hurricanes form over warm tropical seas. They are classified on a scale from 1 to 5, with 5 being the most powerful storm. Katrina was category 4 as it approached the Gulf Coast of the United States.

Most hurricanes that hit the United States start as tropical storms in the Atlantic Ocean. They form during late summer and early fall when the ocean is warmest. As a tropical storm moves west, it draws energy from the warm ocean water. The storm gets larger and stronger, and the wind spins faster and faster.

The spinning wind draws a lot of warm water vapor high in the storm system. When the vapor cools, it condenses. Condensation releases even more energy, which makes the system spin even faster. When the hurricane reaches land, the winds are blowing at deadly speeds, up to 250 km per hour. The rain is very heavy. The wind and rain can cause a lot of destruction.

As soon as a hurricane moves over land, it begins to lose strength. It no longer has warm water to give it energy and water vapor. Within hours, the wind and rain drop to safe levels.

Thunderstorms

Thunderstorms form when an air mass at the ground is much warmer and more humid than the air above. Rapid convection begins. As the warm, humid air rises, the water vapor in it condenses. The condensing water vapor transfers energy to the surrounding air, causing the air to rise even higher. The rapid movement of air also creates a static electric charge in the clouds. When the static electricity discharges, lightning travels from the clouds to the ground, and you hear the sound of thunder. Thunderstorms can cause death, start fires, and destroy communications systems. The powerful winds and heavy rain can cause property damage.

Thunderstorms are most common over land during the afternoon. The Sun heats Earth's surface, and heat transfers to the air. When cold air flows under the warm, moist air, thunderstorms are possible.

Lightning travels from the clouds to the ground.

A tornado spinning through a city

Tornadoes

Tornadoes are powerful forms of wind. They most often happen in late afternoons in spring or summer. When cold air over the land runs into a mass of warm air, the warm air is forced upward violently. At the same time, cooler, more-dense air flows in from the sides and twists the rising warm air. A spinning funnel forms. It "sucks up" everything in its path like a giant vacuum cleaner. The air pressure inside the funnel is very low. The air pressure outside the funnel is much higher. The extreme difference in air pressure can create wind speeds of 400 km per hour or more. Tornadoes can seriously damage everything in their path.

Tornadoes are most common in the south central part of the United States, from Texas to Nebraska. Hundreds of tornadoes occur in this region each year. Warm, moist air from the Gulf of Mexico moves northward. It runs into cooler, drier air flowing down from Canada. This creates perfect conditions for tornadoes. That's why this part of the United States is called Tornado Alley.

A tornado over water is called a waterspout.

Hot and Cold

Death Valley is one of the hottest places on Earth.

Hot and cold weather are the direct result of solar energy. It gets hot when energy from the Sun is intense. It gets cold when solar energy is low. The ocean also affects temperature. The highest and lowest temperatures are never close to the ocean. Water has the ability to absorb and release large amounts of energy without changing temperature much. This keeps places close to the ocean from getting really hot or really cold.

Here is a table of temperature extremes for the United States and the world. These temperatures are deadly for most organisms. Only a few tough organisms are able to survive such temperatures.

Area	Location	High Temperature	Low Temperature
United States	Death Valley, California	57°C	
	Prospect Creek, Alaska		–62°C
World	Death Valley, California	57°C	
	Vostok, Antarctica		–89°C

Weather Extremes

The West Coast and Northeast region of North America do not have many hurricanes or tornadoes. But they do have weather extremes. Most of them involve the ocean.

During the winter, it often rains and snows along the East and West Coasts and in the western mountains. When large storms come in from the Atlantic or Pacific Ocean, wind and rain can cause property damage and flooding. In the mountains, the precipitation comes down as snow. Intense snowstorms are called **blizzards**. A single blizzard can drop 4 meters (m) or more of snow. The snow for a whole winter might exceed 10 m.

A blizzard can drop more than 4 meters (m) of snow.

An ice storm can cause a lot of damage.

The Pineapple Express is a band of warm, moist air that flows to the West Coast from the warm ocean around the Hawaiian Islands. When the warm, humid Pineapple Express meets cold air flowing down from Alaska, a violent winter storm can develop. High winds and heavy rain can uproot trees, destroy homes, and flood large areas of lowlands.

When seasonal rain and snow fail to develop, droughts can occur. A drought is less-than-normal precipitation. In the Southwest, this means less rain in the deserts and hills, and less snow in the mountains. Less snow means less spring runoff. Less runoff means less flow in rivers and streams. Lakes and ponds shrink and in some cases dry up completely. Soil moisture dries up, and ground water decreases. Reservoirs that people use to store water shrink.

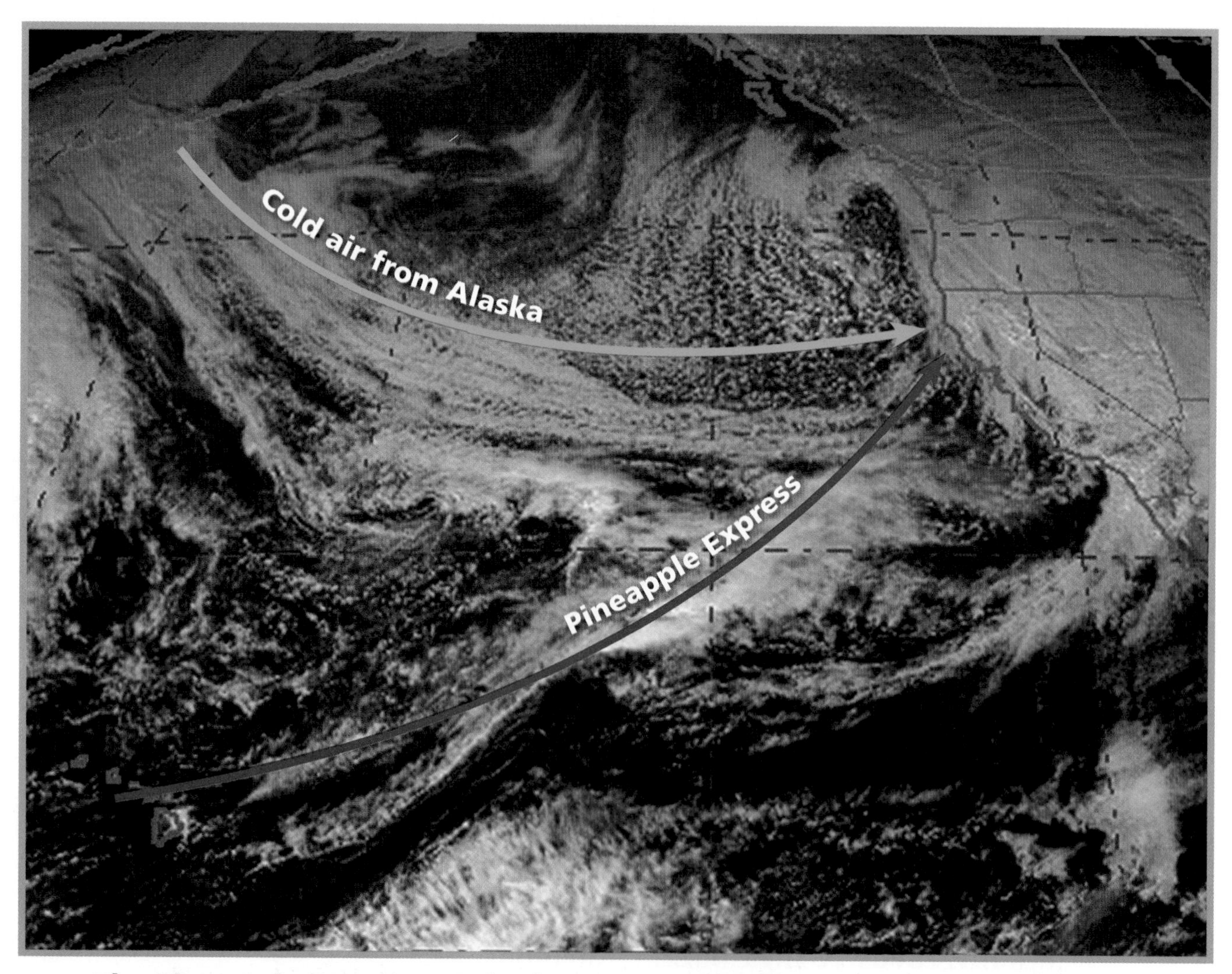

The Pineapple Express carries large amounts of moisture to California.

Water from lakes and rivers dries up during a drought.

Droughts put stress on natural and human communities. Fish and other aquatic organisms might die. Plants that are not adapted for dry environments might die. Reduced water for crops means less food production. People have to conserve water by using less and recycling water when possible.

Serious droughts are not uncommon. During the early 1930s, parts of Colorado, Kansas, New Mexico, Oklahoma, and Texas received little rain. Crops failed. Then came the strong winds. The farms in the area were stripped of their rich topsoil. The farmers had to leave the area because their fields were destroyed. Thousands of families had to leave the area known as the Dust Bowl.

Could it happen again? Many climate scientists think it is happening again now. The precipitation in the Southwest has been declining since the early years of this century. Stream flow and ground water are reduced. Reservoirs are low. The drought that has settled over the Southwest could be part of the overall change in the worldwide **climate**. People in the Southwest should be prepared to use less water. And they should be aware that a general drying of the land could result in more and hotter wildfires.

The Role of the Ocean in Weather

The ocean affects the weather.

The ocean affects weather in the United States in several ways. The ocean is the source of most of the precipitation that falls on the West Coast states. Water evaporates from the ocean, particularly where the Sun has warmed the ocean's surface. Wind carries the water vapor and clouds over the land. As the moist air rises and cools over the coastal mountains, the Sierra Nevada, and the Cascade Range, the water vapor condenses and falls back to Earth's surface. During the spring and summer, the water flows back to the ocean, to complete the water cycle.

The ocean creates mild temperatures all year along the West Coast. It rarely gets too hot or too cold. The temperature of the ocean doesn't change quickly. So the ocean keeps the air temperature near the coast even all year.

The ocean creates breezes near the coast. Because water heats up and cools down slowly, there is often a difference in the temperature of the land and the ocean. Uneven heating starts a convection current, which results in wind. The Sun and the ocean are responsible for ocean breezes.

Thinking about Severe Weather

1. What causes hurricanes?
2. What causes tornadoes?
3. How does the water cycle affect weather along the West Coast?
4. How does the ocean influence the weather along the West Coast?

Earth's Climates

What's the weather like today? What was it like last year on this same date? Probably just about the same. We can guess what the weather will be like tomorrow and next year at this time because weather tends to follow predictable patterns over long periods of time. The big patterns of weather define a region's climate. Climate describes the average or typical weather conditions in a region of the world. The climate in Hawaii is quite different from the climate in Wisconsin. The Hawaiian climate is warm, sunny, and pleasant all year long. The Wisconsin climate is freezing cold in the winter, and hot and humid during the summer.

There are about 12 general climate zones in North America. The two variables that are most important for determining a climate zone are the average temperature throughout the year and the amount of precipitation throughout the year.

This climate map shows the distribution of climate types in North America.

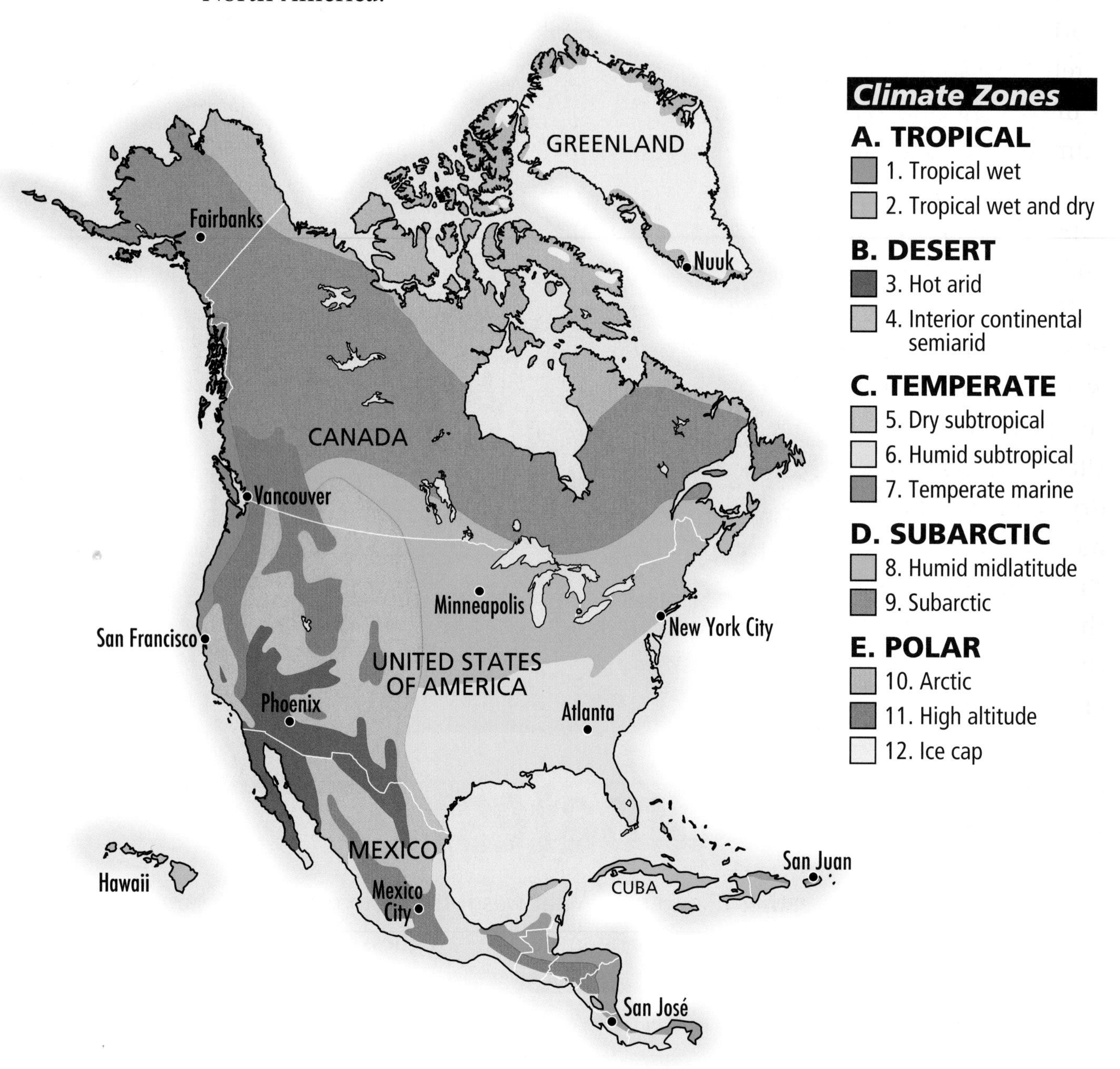

In the Midwest, you can be fairly sure that it will be cold and snowy in January and February and rainy during the summer each year. The same kind of weather will be experienced in Minnesota, Illinois, Connecticut, and Maine. The humid midlatitude climate zone includes the midwestern United States, New England, and the southern part of Canada. This climate zone supports huge diverse forests of deciduous and evergreen trees and all the animals that forests support.

The weather in the southeastern United States is significantly different. Florida, Mississippi, and Louisiana rarely have snow in the winter, and the summers and springs are rainy, hot, and humid. The southern states fall into the humid subtropical climate zone. This zone supports large hardwood forests and many kind of vines.

Humid subtropical zone

The hot arid climate zone in the western United States has predictably warm, dry winters and very hot, dry summers. Arizona and parts of Nevada, Utah, and California are sunny and dry all year. Little rain falls during most of the year. During the summer the temperature can be very high, and thunderstorms can deliver heavy rains that can cause flash floods. The hot arid zone supports a wide diversity of drought-resistant plants, including cactus, mesquite, and yucca, and a host of burrowing and sun-loving animals.

Hot arid zone

Four other climate zones occur in the west (see the climate map). The interior continental semiarid zone is characterized by warm spring and summer weather, cold winters, and summer thunderstorms with the possibility of tornadoes. The semiarid climate supports large expanses of sagebrush and huge grasslands. Land in the interior continental semiarid zone is often used by ranchers to graze livestock.

Interior continental semiarid zone

The high-altitude zone found high in the mountains supports large forests of evergreen trees and provides the right conditions for skiing and other winter sports requiring snow.

Weather in the dry subtropical zone is usually warm and rainy in the winter but hot and dry in the summer. The dry subtropical zone supports oak woodlands, chaparral, and a very diverse community of brush, grasses, and mixed forests. Dry subtropical climates are excellent for farming, fruit orchards, vegetable gardens, and raising livestock.

High-altitude zone

Dry subtropical zone

The temperate marine zone of the Pacific Northwest is cool and wet throughout the year. The climate is strongly influenced by the Pacific Ocean, which keeps the weather cool and moist. This climate zone is characterized by dense forests of large evergreen trees: redwood, fir, pine, and spruce. The moist forests are often home to ferns, mosses, lichens, and fungi. Winters are cool and rainy, while summers are cool and can be foggy and wet.

Two climate zones occur in Alaska. They are the subarctic and the arctic. The climate is extremely cold most of the year, with variable precipitation.

Hawaii has a tropical wet and dry climate, warm and sunny all year long with plenty of tropical rain in many parts of the islands.

Climates vary widely across the country. Many states have only one kind of climate throughout, such as Michigan, Massachusetts, Alabama, and Florida. Other states have two or more kinds of climate. Look at Texas and California. How many climate zones do these states have? So when you are asked what the weather will be like in California, you have to know what part of the state, and what time of year.

Temperate marine zone

Arctic zone

Tropical wet zone

Global Climate Change

Weather is the condition of the atmosphere in a particular place on Earth. Temperature, humidity, and wind describe the weather. The weather is different all over Earth, depending on where you are and the time of year. Climate is the average weather over many years in a region on Earth. The climate in Barrow, Alaska, is very different than the climate in Tahiti. Barrow's climate is extremely cold. The climate on the tropical island of Tahiti is very sunny and warm. Earth's climates are predictable today, but climates have changed many times throughout Earth's history.

Factors Affecting Climate through History

Temperature on Earth is affected by one major energy source, the Sun. The amount of solar energy (heat and light) given off by the Sun is steady day after day, and year after year. But there are many factors that affect the amount of solar energy that transfers to Earth.

An arctic climate

A tropical island climate

Ash from a volcano pollutes the air.

One factor that affects the amount of solar energy transferred to Earth is the amount of pollutants in the air. At times in Earth's history, volcanic eruptions, smoke from forest fires, and major impacts by asteroids and comets have put a lot of dust and many gases into the air. These pollutants act like a shield. They block solar energy from reaching Earth's surface. Smoke and dust that block solar energy can cool the climate in large regions of Earth.

The **greenhouse effect** is another factor that affects the amount of solar energy transferred to Earth. Carbon dioxide, methane, nitrous oxide, and water vapor are greenhouse gases. Greenhouse gases in the air act like a mirror. They let through solar energy, which Earth's surface absorbs. Then Earth's surface transfers the energy to the atmosphere. Once the solar energy is in the atmosphere, the greenhouse gases prevent the energy from easily escaping back into space. Heat builds up and the temperature of the atmosphere rises. The trapping of energy in the atmosphere is the greenhouse effect. It impacts climate worldwide.

The Greenhouse Effect

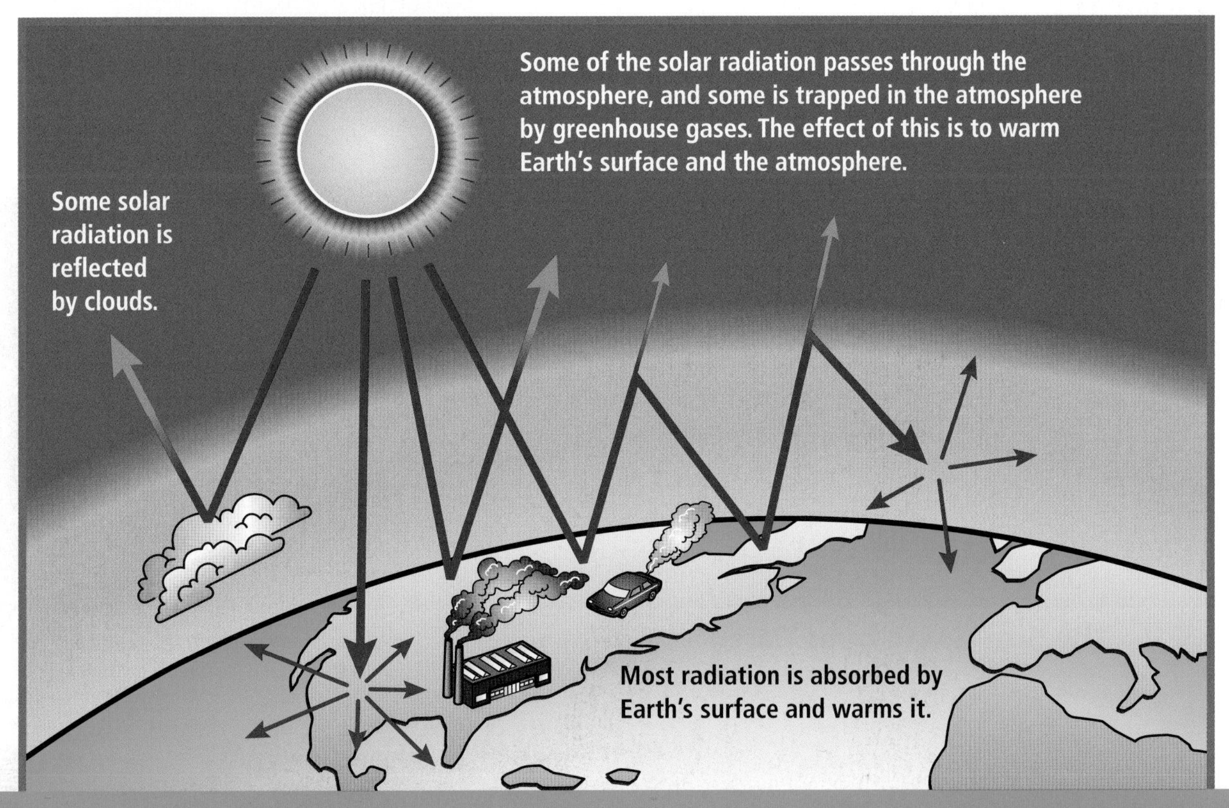

What Is Changing Today?

Today, we are experiencing a period of rapid climate change. The average temperature worldwide has increased about 0.8 degrees Celsius (°C) since 1850. That doesn't sound like much. But think about how much energy it would take to warm all of Earth's atmosphere and the ocean that much. That's a lot of energy.

Scientists have developed climate models that suggest that the global temperature may increase by another 1.5°C–5°C by the year 2100. This temperature change will affect life on Earth and the global climate. Here's how Earth's climate may change.

The Sun rising over Earth

Rising sea level will flood coastal cities.

More farming land will change into desert.

Higher global temperature will cause glaciers and ice sheets to melt worldwide. The arctic polar ice pack will melt completely. Large areas of ice will melt from Greenland and Antarctica. All that melted ice will flow into the ocean and cause the sea level to rise. By 2100, sea level is expected to rise 0.6 meters (m) or more. That is enough to flood many of the low-lying regions of the world. Large parts of Florida, a number of small islands in the Pacific Ocean, and the city of Venice, Italy, would be under water.

As Earth gets warmer, more land will change into desert. Land now used for farming in north central Africa and central Asia will become desert. As a result, the world will produce less food. Food may become more scarce in many parts of the world.

Too much carbon dioxide in the air causes climate change.

What Is Causing Climate Change?

Scientists agree that the major cause of climate change is human activity. The burning of fossil fuels is the number-one human activity affecting climate change. Fossil fuels are the remains of organisms that lived long ago. Over time, those remains changed into oil, coal, and natural gas. When fossil fuels burn, they release carbon dioxide into the air. Humans burn fossil fuels to generate electricity and to power cars.

Carbon dioxide is found naturally in air. In fact, it is essential to life on Earth. Plants use carbon dioxide in the air to produce food by photosynthesis. But since the Industrial Revolution in the 19th century, humans have been releasing more carbon dioxide into the atmosphere than all the plants in the world can absorb. The result is a lot of carbon dioxide in the air. Carbon dioxide is a greenhouse gas. The more carbon dioxide gas we put into the atmosphere, the faster Earth's temperature will rise.

What Can We Do?

So how can we slow climate change? The best way is to stop adding carbon dioxide to the air. But individuals can't make that global decision alone. Governments and big companies need to get involved. For example, the decision to stop burning coal to produce electricity affects many people, because there would be no electricity in their homes. Before we can stop burning fossil fuels to generate electricity, we need to create alternative sources of electricity. Here are some alternative sources for producing electricity.

Wind turbines (windmills) change wind energy into electricity.

Solar energy generates electricity directly with sunlight and solar cells. Solar energy can also produce electricity indirectly by using a large number of curved mirrors as solar collectors. The mirrors focus the Sun's energy on water flowing through a tube. The heated water produces steam that turns electric generators.

A wind farm

A group of solar cells

A system of solar collectors using mirrors

Geothermal energy produces electricity with hot water and steam from volcanic vents to turn generators.

Hydroelectric generators use water flowing through turbines to produce electricity.

Thermonuclear reactions in nuclear power plants create heat to produce steam. The steam turns electric generators to produce electricity.

These alternative sources of electricity are carbon-free. They don't release any carbon into the air. But each alternative source of electricity has its own challenges. If we are going to provide alternative energy to large numbers of homes, it needs to be reliable, safe, and accessible to everyone. We need to develop the technology to convert these primary energy sources (wind, moving water, sunlight, and volcanic vents) into electricity that can be used on a large scale.

A geothermal power plant

A hydroelectric power plant

A nuclear power plant

You can help conserve energy by riding a bike instead of riding in a car.

What Can You Do Right Now?

The most important thing you can do to slow climate change is to conserve energy. Energy used in your home and for transportation is where most of the carbon dioxide in the air comes from. So the next time you need to get from one place to another, don't ask for a ride in the car. Instead, maybe you can walk, or ride your bike or skateboard. Use your own energy instead of carbon energy to get around. And you will get some exercise, too!

At home, you can replace burned-out lightbulbs with compact fluorescent lightbulbs. These use a lot less electricity to produce the same amount of light. If your family needs a new appliance, like a refrigerator, freezer, or water heater, suggest an energy-efficient appliance. You can also adjust the thermostat in your home so it is not too cool or too warm. In the winter, put on a sweater instead of turning up the heat. In the summer, wear lightweight clothes, and enjoy the warmth.

What other ways can you conserve energy? Become an energy detective at school. Find ways your school can conserve energy.

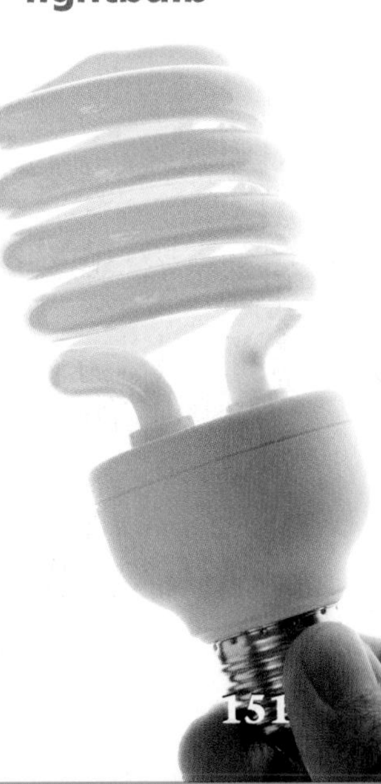

A compact fluorescent lightbulb

Science Safety Rules

1. Listen carefully to your teacher's instructions. Follow all directions. Ask questions if you don't know what to do.
2. Tell your teacher if you have any allergies.
3. Never put any materials in your mouth. Do not taste anything unless your teacher tells you to do so.
4. Never smell any unknown material. If your teacher tells you to smell something, wave your hand over the material to bring the smell toward your nose.
5. Do not touch your face, mouth, ears, eyes, or nose while working with chemicals, plants, or animals.
6. Always protect your eyes. Wear safety goggles when necessary. Tell your teacher if you wear contact lenses.
7. Always wash your hands with soap and warm water after handling chemicals, plants, or animals.
8. Never mix any chemicals unless your teacher tells you to do so.
9. Report all spills, accidents, and injuries to your teacher.
10. Treat animals with respect, caution, and consideration.
11. Clean up your work space after each investigation.
12. Act responsibly during all science activities.

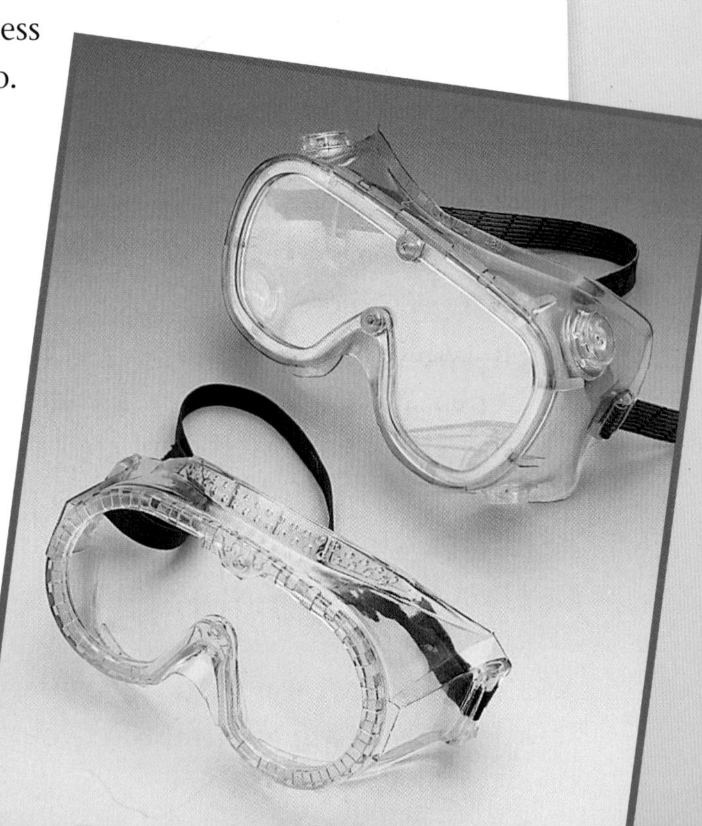

Glossary

absorb to soak in

air the mixture of gases surrounding Earth

air pressure the force exerted on a surface by the mass of the air around or in contact with it

anemometer a weather instrument that measures wind speed with wind-catching cups

asteroid a small, solid object that orbits the Sun

astronomer a scientist who studies objects in the universe, including stars, planets, and moons

astronomy the study of the universe and the objects in it

atmosphere the layer of gases surrounding Earth. The layers include the troposphere, stratosphere, mesosphere, thermosphere, and exosphere.

axis an imaginary line around which a mass, like a planet, rotates

barometer a weather instrument that measures air pressure

Big Dipper a group of seven bright stars in the shape of a dipper

black hole a region in space without light that has a strong gravitational pull

blizzard a severe storm with low temperatures, strong winds, and large quantities of snow

climate the average or typical weather conditions in a region of the world

cloud a large accumulation of tiny droplets of water, usually high in the air

comet a mass of ice, rock, and gas orbiting the Sun

condensation the process by which water vapor changes into liquid water, usually on a surface

conduction the transfer of energy from one object to another by contact

conserve to use wisely and protect carefully

constellation a group of stars in a pattern given a name by people

convection current a circular movement of fluid (such as air) that is the result of uneven heating of the fluid

crater a hole formed by an object hitting a surface

crescent Moon the curved shape of the visible part of the Moon before and after a new Moon

cycle a set of events or actions that repeat in a predictable pattern

day the time between sunrise and sunset on Earth

dew water that condenses outdoors on a surface when the temperature drops at night

diameter the straight-line distance through the center of an object, one side to the other side

drought a less-than-normal amount of rain or snow over a period of time

dwarf planet a round object that orbits the Sun and does not orbit a planet

Earth the third planet from the Sun, known as the water planet

energy transfer the movement of energy from one place to another

evaporation the process by which liquid water changes into water vapor

exosphere the layer of the atmosphere above the thermosphere. The exosphere is the transition from the atmosphere to space.

extraterrestrial beyond Earth

first-quarter Moon a phase of the Moon in the lunar cycle halfway between a new Moon and a full Moon

fog a large accumulation of water droplets close to the ground

force a push or a pull that acts on an object or a system

fossil fuel the preserved remains of organisms that lived long ago and changed into oil, coal, or natural gas

frost frozen condensation on a surface

full Moon the phase of the Moon when all of the sunlit side of the Moon is visible from Earth

galaxy a group of billions of stars. Earth is in the Milky Way galaxy.

gas a state of matter with no definite shape or volume; usually invisible

gas giant planet one of the four planets that are made of gas. These are Jupiter, Saturn, Uranus, and Neptune.

geosphere the solid, rocky part of Earth's crust

gibbous Moon the shape of the Moon when it appears to be more than a quarter but not yet full and when it is less than full but not quite a third quarter.

gravitational attraction the mutual force pulling together all objects that have mass

gravity the force of attraction between masses

greenhouse effect the trapping of heat in the atmosphere causing Earth's temperature to rise

hail precipitation in the form of balls or pellets of ice

humidity water vapor in the air

hurricane a severe cyclonic tropical storm that produces high winds and heavy rainfall

hydrosphere all of the water on Earth in solid (ice), liquid (water), and gas (water vapor) phase

hygrometer a weather instrument that measures humidity

kinetic energy energy of motion

Kuiper Belt a huge region beyond the gas giant planets, made up of different-sized icy chunks of matter

liquid a fluid state of matter with no definite shape but definite volume

lunar cycle the 4-week period during which the Moon orbits Earth one time and is visible in all of its phases

lunar eclipse the effect observed when Earth passes exactly between the Moon and the Sun, casting its shadow on the full Moon

magnify to make an object appear larger

mass the amount of material in something

matter anything that takes up space and has mass

mesosphere the region of the atmosphere above the stratosphere

meteorologist a scientist who studies the weather

Milky Way the galaxy in which the solar system resides

Moon Earth's natural satellite

new Moon the phase of the Moon when the sunlit side of the Moon is not visible from Earth

night the time between sunset and sunrise on Earth

observatory a building that houses a telescope

opaque matter through which light cannot travel

orbit to move or travel around an object in a curved path (can also be a noun, the path traveled)

ozone a form of oxygen that forms a thin protective layer in the stratosphere

phase the shape of the visible part of the Moon

photosynthesis a process used by plants and algae to make sugar (food) out of light, carbon dioxide, and water

planet a large, round object orbiting a star

planetarium a theater with a dome-shaped ceiling that represents the sky

precipitation rain, snow, sleet, or hail that falls from clouds

predict to estimate accurately in advance based on a pattern or previous knowledge

radiant energy energy that travels through air and space as waves

radiation energy that travels through air and space as waves

rain liquid water that is condensed from water vapor in clouds and falls to Earth in drops

recycle to use again

reflect to bounce off an object or surface

renewable resource a natural resource that can replenish itself naturally over time. Air, plants, water, sunlight, and animals are renewable resources.

rotate to turn on an axis

satellite an object, such as a moon, that orbits another object, such as a planet

season a time of year that brings predictable weather conditions to a region on Earth

severe weather out-of-the-ordinary and extreme weather conditions

shadow the dark area behind an object that blocks light

sleet precipitation in the form of ice pellets created when rain freezes as it falls to Earth from the atmosphere

snow precipitation in the form of ice crystals (snowflakes)

solar eclipse the visual effects created when the Moon passes exactly between Earth and the Sun

solar energy heat and light from sunshine

solar system the Sun and the eight planets and other objects that orbit the Sun

solar wind the steady flow of particles from the Sun

star a huge sphere of hydrogen and helium gas that radiates heat and light

stratosphere the region of the atmosphere beyond the troposphere. The ozone layer is in the stratosphere.

Sun the star at the center of the solar system around which all of the solar system objects orbit

sunrise a time in the morning when the Sun appears over the horizon. The Sun always rises in the east.

sunset a time in the evening when the Sun disappears under the horizon. The Sun always sets in the west.

telescope an optical instrument that makes distant objects appear closer and larger

temperature a measure of how hot something is

terrestrial planet the four small, rocky planets closest to the Sun. These are Mercury, Venus, Earth, and Mars.

thermometer a weather instrument that measures temperature

thermonuclear reaction a change in atomic structure that creates huge amounts of heat and light energy, such as the reactions that occur in the Sun

thermosphere the region of the atmosphere above the mesosphere

third-quarter Moon a phase of the Moon in the lunar cycle halfway between the full Moon and the new Moon

thunderstorm severe weather that produces heavy rainfall and powerful static electric discharges

tornado a rapidly rotating column of air that extends from a thunderstorm to the ground. Wind speeds can reach 400 kilometers (km) per hour or more in a tornado.

transpiration the process by which water escapes into the air from plants

troposphere the region of the atmosphere that begins at Earth's surface and extends upward for 9 to 20 km. Weather happens in the troposphere.

unaided eyes looking at something without the use of a telescope or microscope

waning getting smaller

water cycle the global water-recycling system. Water evaporates from Earth's surface, goes into the atmosphere, and condenses and returns to Earth's surface as precipitation in a new location

water vapor the gaseous state of water

waxing getting larger

weather the condition of the air around us. Heat, moisture, pressure, and movement are three important variables that determine weather.

weather variables data that meteorologists measure. These include temperature, wind speed and direction, air pressure, cloud cover, and precipitation.

wind air in motion

wind meter a weather instrument that measures wind speed

wind vane a weather instrument that indicates wind direction

Index